T5-AFM-282

Tropical Forestry

Florencia Montagnini · Carl F. Jordan

Tropical Forest Ecology

The Basis for Conservation and Management

With 56 Figures and 24 Tables

Dr. Florencia Montagnini
Professor in the Practice of Tropical Forestry
Director, Program in Tropical Forestry
Yale University
School of Forestry and Environmental Studies
370 Prospect St.
New Haven, CT 06511
USA

Dr. Carl F. Jordan
Senior Ecologist
Institute of Ecology
University of Georgia
Athens, GA 30602-2202
USA

ISSN 1614-9785
ISBN 3-540-23797-6 Springer Berlin Heidelberg New York

Library of Congress Control Number: 2004116536

Springer is a part of Springer Science+Business Media

springeronline.com

Printed in The Netherlands

Editor: Dr. Dieter Czeschlik, Heidelberg, Germany
Desk Editor: Anette Lindqvist, Heidelberg, Germany
Production: PRO Edit GmbH, Elke Beul-Göhringer, Heidelberg, Germany
Typesetting: K+V Fotosatz GmbH, Beerfelden
Cover design: design & production GmbH, Heidelberg, Germany

31/3150/beu-göh – 5 4 3 2 1 0 – Printed on acid-free paper

Preface

In 1973, a group of tropical ecologists gathered at Turrialba, Costa Rica, for a workshop to assess the knowledge of tropical forest ecology, and to make recommendations for future study. The proceedings were published in a volume entitled *Fragile Ecosystems* (Farnworth and Golley 1974). The book was called *Fragile Ecosystems* because many ecologists with experience in low latitudes suspected that tropical forests, especially rain forests, were particularly susceptible to disturbance. Recovery following activities such as logging and shifting cultivation was thought to be slower and more difficult than recovery of temperate zone forests. If, in fact, this were the case, it would have important implications for management of tropical forests. However, at the time, there was very little evidence that tropical forests were especially "fragile".

In the intervening years, hundreds if not thousands of studies were published on rain forest ecology. Many have bearing on the question of whether tropical forests are more easily damaged than temperate forests and, if so, why. This question is particularly important for forest management, since tropical forest management is often carried out with methods developed for temperate zone forests. The purpose of this book is to bring together evidence that bears on the question of the uniqueness of tropical ecosystems, and to examine what this evidence means for the management of tropical forests in a way that does not diminish the ecosystem's ability to maintain its structure and function.

Chapter 1 of this book reviews the values of tropical forests, both commercial and non-market values, that will disappear if tropical forests become extinct. To ensure that these values are not lost, we must make sure that tropical forests themselves are perpetuated.

In order to develop approaches to forest management that will promote forest survival, it is necessary to understand the characteristics of tropical forests that are important for maintaining their structure and function. Especially important is how tropical forests differ from temperate forests, since forest management techniques developed in the temperate zone may not be appropriate for the tropics. Chapter 2 describes these ecological characteristics.

Chapter 3 reviews several schemes of classification. Classification of tropical forests can be important in determining management plans. There are many ways to classify tropical forests, but most are based either on climate or on stand structure. The problem is that within one climatic zone, there can be a variety of forest functions. Also, forests with similar structures can function differently. Because function is not taken into account, many traditional classification schemes are not useful at the stand level. Chapter 3 proposes other approaches that may complement the traditional classifications.

Social and economic factors usually play a more important role in management decisions than do ecological factors, and it is the social and economic pressures that are driving tropical deforestation. In Chapter 4 we examine the proximate and the underlying causes of deforestation, and its effects on the environment and on human populations.

Chapter 5 shows how the understanding of tropical forest ecology, together with considerations of local economy and culture can be applied to sustainable forest management. Methods of forest management are discussed, along with their effects on biodiversity.

Chapter 6 examines the multiple roles of plantation forestry: production of timber and fuelwood; a tool for development; and preserving or recovering biodiversity. Agroforestry systems are also put forward as an alternative to reconcile production with conservation and social needs. Plantation forestry, agroforestry, and other techniques are also presented as tools to aid in restoration of degraded forests and degraded agricultural and pasture lands.

Chapter 7 contrasts the impact of decisions made at the regional, national, and international levels with those made locally on sustainability of the forest. The top-down approach to development is contrasted with bottom-up approaches. Case studies where community forestry has been successful at implementing sustainable forest management are presented. Finally, Chapter 8 synthesizes what we have learned, and how that knowledge can be applied to future management decisions.

F. Montagnini
C.F. Jordan

Contents

Importance of Tropical Forests 1

1.1 Functions of Tropical Forests

The functions of tropical forests can be productive (timber, fiber, fuelwood, and non-timber forest products), environmental (climate regulation, carbon sequestration and storage, reserve of biodiversity, and soil and water conservation), and social (subsistence for local populations and cultures). Forests serve a combination of functions and can generate additional revenue for local populations and national economies through ecotourism. Forests also have aesthetic, scientific, and religious values. In this chapter, we examine the principal productive and environmental services of tropical forests.

1.2 Economic

1.2.1 Forest Products

1.2.1.1 Timber

Wood is one of the most useful and versatile raw materials. Compared to most available materials, wood is stronger, more workable, and more aesthetically pleasing (Wadsworth 1997). Wood is also warm to the touch, which makes it preferable for flooring and other house construction purposes. In addition, wood products are biodegradable, which is an added environmental advantage.

Commercial timber production is a major global industry. In 1998, global production of industrial roundwood (all wood not used as fuelwood) was 1.5 billion m^3 (FAO 2000 a). In the early 1990s, production and manufacture of industrial wood products contributed about US$ 400 billion to the global economy, or about 2% of global GDP (World Resources Institute 2000).

Although the roundwood timber market is dominated by North America and Europe, the timber industry is of greater economic importance to developing countries such as Cambodia, the Solomon Islands, and Myanmar, where wood exports account for 30% of all international trade. On average, timber constitutes about 4% of the economies of developing countries (Myers 1996).

Furthermore, the global demand for timber is expected to increase over the next decade. There have been signs of scarcity in some of the more precious woods. Production of tropical wood products has recently fallen below earlier levels, and some Asian countries have experienced difficulties in reaching their expected volumes of exports (FAO 2001 b). Forest industries continue to adapt to changes in raw materials, namely the increased supply of plantation wood from a wider variety of species (Figs. 1.1 and 1.2).

For most developing nations, there is a lack of reliable data on net annual forest growth, removal rates, and the age of trees – information that is needed to accurately assess the long-term conditions of forests. Even so, there is considerable evidence that, in some regions, harvest rates greatly exceed regrowth (World Resources Institute 2000). Certain valued species such as mahogany (*Swietenia macrophylla*) and teak (*Tectona grandis*) are harvested at rates that will eventually lead to depletion of these species from the forest. For example, in Thailand, forest cover diminished from 53 to 28% between 1961 and 1988, with much of the loss in the teak forests of the north (Phothitai 1992). In response, private industry initiated a teak reforestation program.

Fig. 1.1. Timber scarcity has led to the utilization of smaller diameters and shorter logs in many tropical regions. These logs were extracted from natural forests for their use for furniture in Petén, Guatemala (Photo: F. Montagnini)

Fig. 1.2. In many developing countries timber exports are not just roundwood but processed timbers as in fine furniture. In this furniture factory in Guatemala they manufacture house furniture for export to retailers in the USA (Photo: F. Montagnini)

1.2.1.2 Fuelwood

Fuelwood, charcoal, and other wood-derived fuels (collectively known as woodfuels) are the most important form of non-fossil fuel. The world production of fuelwood for 1999 was about 1,700 million m^3, of which roughly 90% was produced and consumed in developing countries (FAO 2001a). Biomass energy, which includes woodfuels, agricultural residues, and animal wastes, provides nearly 30% of the total primary energy supply in developing countries. More than 2 billion people depend directly on biomass fuels as their primary or sole source of energy.

In developing countries, woodfuels account for more than half the biomass energy consumption (World Resources Institute 2000). At least half the total timber cut in these countries is used as fuel for cooking and heating. Scarcity is more acute in the Indian subcontinent and in semiarid regions of Africa below the Sahel. In Latin America, firewood scarcity is a problem in the Andean region, Central America, and the Caribbean (Fig. 1.3). Whether a regional or even global fuelwood crisis will develop depends on a variety of factors, such as the increase in the area of plantations for fuelwood, the use of more efficient burning stoves, and the availability of alternative sources of

Fig. 1.3. Fuelwood scarcity is a serious problem in several rural areas of Central America, especially in the drier forest regions. This oxen cart is attempting to cross a river on its way to the market in Jinotepe, near the Pacific coast of Nicaragua (Photo: F. Montagnini)

energy. However, there is little doubt that growing fuelwood scarcity will increase the economic burden on the poor in some regions.

1.2.1.3 Non-Timber Forest Products

Non-timber forest products (NTFPs) include a myriad of products that have been extracted from forests around the world for millennia (Table 1.1). Because most of these products are consumed locally they used to be called "minor forest products," disregarding their importance. In many countries NTFP extraction can be a major economic activity. For example, in India NTFPs are a critical component of the economic activity of about 500 million people living in or around forests. In the 1970s, the economic value of NTFPs in India surpassed that of timber sales, and currently export earnings from NTFPs account for 60–70% of total exports from forest products (Thadani 2001).

Rattans and Bamboos

Rattans and bamboos are economically the most important NTFPs in Asia. As rattan requires arboreal support and shade, its cultivation does not re-

Table 1.1. Examples of non-timber forest products (NTFPs), extracted and collected by local people from tropical forests

Product	Examples
Fuel and fodder biomass	Leaves, branches, roots of trees, and shrubs
Construction materials	Canes, rattan, bamboo, palms
Fiber	Palms, lianas, herbs
Flowers	Orchids, anthurium, passion flower
Ornamental plants	*Zamia*, *Chamaedoera*
Fruits	Zapote, durian, Brazil nut, açai
Tubers	Yams, taro
Other edible plant parts	Heart of palm
Mushrooms	Variety of edible mushrooms
Seeds	Colorful seeds for crafts
Oils	*Dipteryx odorata*, palm oil
Medicinal plants	*Quassia amara*, *Smilax*, *Cinchona*
Gums and resins	Rubber, chicle
Tannins and dyes	Brazil tree (*Caesalpinia echinata*)
Wildlife products	Honey, eggs, feathers, birds, mammals, fish, insects

quire clear felling of forest, but rather requires that forests are left standing. Cultivation of rattan can be eight times more profitable than rice (Thadani 2001). Bamboo is used mainly for furniture but also for paper and food, as the shoots are edible.

Materials for Crafts

Many wood products are used to make crafts that are sold as souvenirs in local and international markets. In Nicaragua, a great variety of crafts are produced from wood from non-conventional species that do not have other uses, as well as from fiber extracted from secondary forests. The markets of Masaya and Masatepe in southern Nicaragua are visited by local and international tourists (Santana et al. 2002). The extraction and marketing of these products have a substantial impact on local economies, involving a variety of people, including the people who extract the NTFPs from the forest, the intermediaries, the artisans, and the sellers in the local markets.

Edible Products

Edible NTFPs not only have economic value in local and international markets, but also provide food security to local populations, especially during periods of drought or famine. Many of the more important edible NTFPs that formerly were collected from the native forests are now cultivated in commercial plantations. Some examples are mangosteen, *Garcinia mangostana*, durian, *Durio zibethinus*, zapote, *Pouteria sapote*, and Brazil nut, *Bertholletia excelsa*. In Costa

Fig. 1.4. A 9-year-old plantation of *Euterpe edulis* palm at a CEPLAC (Center for Cacao Research and Extension) station in Una, Bahia, Brazil
(Photo: F. Montagnini)

Rica, the palm *Bactris gasipaes*, which used to be cut from forest for palm-heart, is now planted commercially in extensive monocultures. The fruit of the açai palm, *Euterpe oleracea*, is an important food for the inhabitants of the floodplain forests in the Amazon estuary near Belem, Pará, Brazil (Muñiz-Miret et al. 1996). The fruits are also sold in the Belem market. Because açai is so profitable, farmers plant it in home gardens and manage natural açai stands by cutting back other plants that may compete with it. The extraction of another palm of the same genera, *Euterpe edulis*, for its palm-heart has led to overexploitation of the species in Brazil and Argentina (Fig. 1.4).

Medicines and Insecticides

Many modern medicines originated in forests around the world. Salicylic acid (a component of aspirin) was first isolated from willows, and quinine (used to treat malaria) was discovered in *Cinchona officinalis*. Much of this

Fig. 1.5. Leaves and branches of *Quassia amara* trees are collected from forests throughout this species' broad range in Central America for its medicinal and insecticidal uses. This picture was taken in the Kéköldi Indigenous Reserve in Talamanca, Costa Rica (Photo: CATIE)

knowledge is discovered through ethnopharmacology, the study of indigenous herbal medicines. Some medicinal plants can also have biocidal properties. For example, extracts of *Quassia amara*, a medicinal tree that grows in forests throughout Central America, have been tested by CATIE (Tropical Agriculture Research and Higher Education Center) in Turrialba, Costa Rica, as an insecticide to control the mahogany shoot borer *Hypsipyla grandella* (Montagnini et al. 2002; Fig. 1.5).

Rubber and Resins

Rubber extracted from *Hevea brasiliense* is an important economic activity for many rubber-tapper communities that live in the Brazilian Amazon. Rubber extraction was recognized as such an important activity for local people that in the 1980s "extractive reserves" were officially designated to protect the forests and ensure the livelihoods of the local people. Chicle from *Manilkara zapota* has long been an important NTFP in Petén, a region in Guatemala that contains many important forest resources.

Ornamental Plants

Many ornamental plants are extracted from forests for commercial purposes. For example, the extraction of the "xate palm", *Chamaedoera* spp., used for floral arrangements, is a very important activity in Petén, Guatemala. *Zamia skinneri*, a Cycadaceae, has been actively extracted from Central American forests for ornamental purposes to such an extent that its populations are endangered in many regions and its extraction has been banned (Montagnini et al. 2002).

NTFPs and Local Populations

In what has become a classic article, Peters et al. (1989) demonstrated the potential economic importance of NTFP to local populations. The article showed for the first time that a hectare of forest in Iquitos, Peru, harvested for NTFPs, could yield a higher economic benefit than other more destructive land uses such as slash-and-burn agriculture and cattle. Their results were limited to their study area and some economists criticized their work because it did not consider possible market saturation if NTFP production increased. However, other authors conducted similar studies in other forests and the idea that, due to the economic importance of NTFPs, forests may be worth more when intact than when exploited was generalized in the 1980s and 1990s.

In developing countries, the dependence of people on NTFPs may be higher than in developed countries. In developing countries, unemployment is often high and unemployed people generally do not receive good government subsidies; thus the extraction and sale of NTFPs can be an important contribution to income generation. Traditional medicines are often the only or principal healing aid in many forest communities. Fruits that are rich in vitamins can be important in the diet of local people. Ornamental plants extracted from forests are used by local people for their aesthetic value. Animals that inhabit the forest are often important sources of protein for local populations. However, overhunting has seriously depleted game populations in some tropical forests.

In general, there are two types of non-market values, attributable and intangible or non-assignable (Farnworth et al. 1981). Fruits, medicines, animals, ornamental plants, and other products can be attributed a market value, while for other NTFPs it is more difficult or impossible to assess a value. Some social, cultural, and religious values are very difficult to quantify. In addition, indigenous people living in forests often have strong religious and cultural links to the forest. For them, the extraction of NTFPs relates to their cultural and religious beliefs.

1.2.2 Ecotourism

Ecotourism represents one of the most environmentally friendly alternatives for the economic development of protected areas (Li and Han 2001). Ecotourism can benefit protected areas by providing income that can make them economically independent and justifying them from a national development perspective (Boza 2001). Ecotourism is a booming business and constitutes a potentially valuable non-extractive use of tropical forests. A major part of non-consumptive recreational activities such as hiking, bird watching, wildlife viewing, and other such pursuits occur within forests.

Ecotourism can be the largest proportion of the tourist industry in a country, as demonstrated in Costa Rica and Belize (Boza 2001). In Costa Rica, tourism is the second largest source of income for the country, bringing in about US$ 900 million a year. In Costa Rica, 1 million tourists visited the country in 2000 and more than half of them visited the forests in either public protected areas or private lands (Nasi et al. 2002). However, many different stakeholders capture the values generated and the profits often leave the country and provide little benefit to local populations. Although the percentage of total value that accrues at the local forest level through ecotourism tends to be small or non-existent, even a minor amount may constitute an important section of the national economy.

The potential of ecotourism varies widely throughout the tropics. Ecotourism may be more feasible in high-quality forests of a fragmented landscape where there is a developed infrastructure and easy access, rather than in large and remote frontier forests. While there is a clear upward trend in global economic revenues from tourism, international tourism is highly sensitive to security problems and political turmoil, causing large fluctuations in income generated by tourism. In addition, if management is poor ecotourism can lead to degradation of the natural resources on which it depends. Thus it is important to evaluate the carrying capacity of protected areas to ensure that they can handle levels of visitation that enable them to become economically and ecologically sustainable (Maldonado and Montagnini 2004).

1.3 Environmental Services

1.3.1 Reserve for Biodiversity

Estimates of numbers of species in tropical forests vary. However, statements that tropical forests harbor a great bulk of the Earth's species are relatively common. For example, Erwin (1988) and Wilson (1992) stated that while covering just 6% of the Earth's land surface, tropical forests are estimated to contain at least 50%, possibly 70%, or even 90% of the Earth's total number of species. It is estimated that about 170,000 plant species, or two-thirds of all plant species on earth, occur in tropical forests (Raven 1988). More recent estimates yield even higher numbers with more than 200,000 species of flowering plants alone (Prance et al. 2000).

Wilson (1992) gives a well-known example of the species richness in tropical forests: a single small tree in the Peruvian Amazon contained as many species of ants as the British Isles. Alwyn Gentry set the world record for tree diversity at a site in the rain forest near Iquitos, Peru. He found about 300 species in each of two 1-ha plots (Wilson 1992). A 1-ha plot in lowland Malaysia contained as many as 250 or more species of trees larger than 10 cm in diameter (Whitmore 1984). Peter Ashton discovered over 1,000 species in a combined census of ten selected 1-ha plots in Borneo (Wilson 1992). Comparisons of number of species between tropical and temperate regions help to illustrate the point: for example, half a square kilometer of Malaysia's forests had as many tree and shrub species as the whole of the USA and Canada (Myers 1996). As forests disappear, so do their species, and most extinctions occurring today happen in tropical forests (Myers 1994).

1.3.2 Regulation of Climate

1.3.2.1 Local Effects

The role of forests and their vegetation in maintaining lower ambient temperatures or higher relative humidity are thought to affect both local and regional climatic conditions (Nobre et al. 1991). This may be important for maintaining or enhancing the productivity of agriculture in adjacent areas (Lopez 1997).

Even more dramatic can be the influence of large extensions of tropical forest on the hydrologic regime. For example, a classic work by Salati and Vose (1984) showed that rainfall in the Amazon is internally recycled, with about 50% of rain coming from condensation of water vapor from evapotranspiration from the forest canopy. When a sizeable amount of forest is cut down, the remaining forest is less able to evaporate and transpire, causing a decrease in rainfall that may eventually result in changes in the vegetation from forest to savanna or woodland (Salati and Nobre 1992). However, the effects of deforestation on rainfall are not that clear. Deforestation also changes the surface albedo and aerodynamic drags, which in turn affects temperatures, cloudiness, air circulation, etc., resulting in a highly scale-dependent and non-linear system (Chomitz and Kumari 1995). Comprehensive reviews of results obtained at different scales using micro-scale empirical studies, meso-scale climate models, and general circulation models show that it is no longer clear whether deforestation reduces rainfall. Some reviews conclude that while the assumption that deforestation affects local climate is plausible and cannot be totally dismissed, the magnitude and outcome of the effect remain to be clearly demonstrated, and are likely to be relatively minor (Nasi et al. 2002).

However, deforestation has been considered responsible for declines in rainfall in several areas of the humid tropics, such as near the Panama Canal, northwestern Costa Rica, southwestern Ivory Coast, western Ghats of India, northwestern Peninsular Malaysia, and parts of the Philippines (Myers 1988; Meher-Homji 1992; Salati and Nobre 1992). In northwestern Peninsular Malaysia, two states have experienced disruption of rainfall regimes, causing 20,000 ha of rice paddy fields to be abandoned and another 72,000 ha to significantly decline in productivity (Myers 1997).

The impact of deforestation on hydrologic flows is a major concern throughout Central America. Sedimentation of reservoirs, dry season water shortages, flooding, and the severity of damage caused by Hurricane Mitch in 1998 have all been widely attributed, at least in part, to deforestation (Pagiola 2002). When vegetation is removed, soil aggregates break down and the soil becomes less permeable to water. As a result, there is less water stored in the soil and more erosion and surface runoff during rainstorms. Consequently, floods are more common during rainstorms and water flow in streams decreases during dry spells.

Both rural and urban populations perceive water services as ecosystem functions that should be maintained. A study carried out by the Tropical Agriculture Research and Higher Education Center, CATIE, in Costa Rica found that most Costa Ricans agree to pay for the environmental services (ES) provided by forests. The same study shows that the ES that Costa Ricans value most is water protection; followed by biodiversity protection, carbon sequestration, and scenic beauty (35, 25, 20, and 20%, respectively) (Nasi et al. 2002).

The poor sectors of human populations worldwide are particularly vulnerable to climate change. Not only are they more dependent on the weather for their livelihoods (for example, through rain-fed agriculture) but also they tend to reside in tropical areas that are likely to suffer the most from rising temperatures and sea levels (Pagiola 2002). Moreover, the poor generally lack the financial and technical resources that could allow them to adjust to global warming.

1.3.2.2 Global Effects

Carbon dioxide (CO_2) is considered to be the principal gas responsible for the "greenhouse effect." One of the most important roles of forests in ameliorating the "greenhouse effect" is in absorbing carbon from the atmosphere, thereby reducing the buildup of carbon dioxide. The effect is as follows: energy in the form of visible light and UV (ultraviolet) rays from the sun passes freely through the atmosphere. It heats up the Earth, but is restricted in its ability to escape when reradiated in the form of infrared radiation because it is absorbed by CO_2 and other gases contained in the Earth's atmosphere. Atmospheric CO_2 increased linearly from 1850–1960, but has increased exponentially since then. Widely dismissed as far-fetched only a few years ago, global warming is currently recognized as real and dangerous (Bishop and Landell-Mills 2002).

Green plants take up CO_2 from the atmosphere and use it in photosynthesis to produce sugar and other plant compounds for growth and metabolism. Long-lived woody plants store carbon in wood or other tissues. Through the process of decomposition the carbon in the wood may be released to the atmosphere after the plants die in the form of CO_2 or methane. As plant material decomposes in the soil, part of the carbon in plant tissue can form part of the soil organic matter, serving as another more or less permanent carbon sink. However, when a forest is cleared for agriculture, pasture, or other purposes, all the carbon stored in the trees and soil is released into the atmosphere.

Due to sampling and measurement problems, measurements of carbon stocks are not very accurate. For the past 20 years, scientists have been attempting to calculate the global carbon stocks of tropical forests, as well as the changes in these stocks as changes in land use occur. Recently, satellite data and remote sensing have been used to characterize ground cover, providing more accurate estimates of changes in vegetation each year (Loveland and Belward 1997).

Globally, forests contain more than half of all terrestrial carbon, and account for about 80% of carbon exchange between terrestrial ecosystems and the atmosphere. Forest ecosystems are estimated to absorb up to 3 billion - tons of carbon annually. In recent years, however, a significant portion of this has been returned through deforestation and forest fires. For example, tropical deforestation in the 1980s is estimated to have accounted for up to a quarter of all carbon emissions from human activities (FAO 2001a).

The carbon stock estimates by the International Panel on Climate Change (IPCC 2000) are listed in Table 1.2. Tropical forests are by far the largest carbon (C) stock in vegetation, while boreal forests represent the largest C stock

Table 1.2. Global carbon stocks in vegetation and soil carbon pools to a depth of 1 m (IPCC 2000)

Biome	Area (10^9 ha)	Global carbon stocks (Gt C)		
		Vegetation	Soil	Total
Tropical forests	1.76	212	216	428
Temperate forests	1.04	59	100	159
Boreal forests	1.37	88	471	559
Tropical savannas	2.25	66	264	330
Temperate grasslands	1.25	9	295	304
Deserts and semideserts	4.55	8	191	199
Tundra	0.95	6	121	127
Wetlands	0.35	15	225	240
Crop lands	1.60	3	128	131

in soils. Tropical savannas store about one third as much C in vegetation as do tropical forests. However, savannas also have large C stocks in soils, similar to those of temperate grasslands. Croplands worldwide have the smallest C stocks in vegetation, with intermediate values for soils.

Because of the importance of forests and forest soils as a sink for carbon, scientists are beginning to agree that forests must be preserved or re-established. Forestry-based carbon sequestration is based on two approaches: active absorption in new vegetation and preservation of existing vegetation. The first approach includes any activity that involves planting new trees (such as afforestation, reforestation, or agroforestry) or increases the growth of existing forests (such as improved silvicultural practices). The second approach involves preventing the release of existing carbon stocks through the prevention or reduction of deforestation and land-use change, or reduction of damage to existing forests. This may involve forest conservation or indirect methods such as increasing the production efficiency of swidden agriculture. Improved logging practices and forest fire prevention are other examples of actions that protect existing carbon stocks.

In 1997, the Kyoto protocol affirmed reforestation and additional incorporation of carbon into agriculture as potential substitutes for reducing the CO_2 emissions from fossil fuels (FAO 2001a). Using reforestation in the tropics as a method for mitigating CO_2 emissions had been a topic of discussion long before 1997. In deciding on the best strategy for addressing these issues, it has been suggested that choices include focusing efforts on protecting primary tropical forests, allowing regrowth of secondary forest in areas that have been cleared, establishing plantations in cleared areas, and encouraging agroforestry on land cleared for agriculture (Cairns and Meganck 1994). Carbon conservation is regarded as having the greatest potential for slowing the rate of climate change. In contrast, carbon sequestration is a slow process.

Recovery of a tropical forest with maximum carbon content can take hundreds of years (Montagnini and Nair 2004).

Some tropical countries have recently started programs of incentives to encourage tree plantation development to help offset C emissions. Since 1966, Costa Rica has contributed payments for environmental services (ES) such as promoting forest conservation, sustainable forest management, and tree plantations through the assignment of differential incentives for each of these systems. Funding for these incentives comes from a special tax on gasoline and from external sources (Campos and Ortíz 1999). In 2003, agroforestry systems were added to the list of systems receiving incentives in Costa Rica, while forest management was eliminated from the list, due to lack of funding to support all incentives and to pressure from environmentalist groups. In several other tropical countries economic incentives are given for the establishment of agroforestry systems in the form of carbon credits (Dixon 1995). For example, the Dutch government is engaged in a 25-year program to finance reforestation projects covering 2,500 km^2 in South America, in order to offset carbon emissions from coal-fired stations in the Netherlands (Myers 1996). As the concept of "carbon credits" being paid by fossil fuel emitters to projects that sequester or reduce carbon output becomes more common, many nations and organizations will seek to find inventive ways to sequester carbon.

1.4 Social

1.4.1 Subsistence for Local Populations

Humans have lived in tropical forests for millennia. The archaeological records for the Niah cave, Sarawak, go back about 40,000 years, and for Amazonia about 5,000–11,000 years (Roosevelt et al. 1996; Whitmore 1998). In Africa, the occupation of the forest has been traced back at least 2,000 years (Wilkie 1988). The ancestors of the Pygmies who now live in the rain forests of the Congo basin were probably the first inhabitants of the African rain forests.

The people of the tropical rain forests of the world are very varied. They differ in their effects on the forest and on the ways in which the forest affects them. Today hunter-gatherer societies still live in all three rain forest regions, living off the wild plants they collect and animals they hunt. However, these are a vanishing minority. Many more people living in the rainforest participate in markets – local, national, or international. They fell trees, plant crops, and make a living from forest resources, thus strongly impacting the forests they dwell in. Also there are those who may not live in the forests themselves but whose lives depend on forests, through managing large forest enterprises or otherwise transforming the forests in a variety of ways. Thus people who rely on the rain forest differ greatly in their understanding of what they are using, destroying, or replacing, and in the value they place on the forest (Denslow and Padoch 1988).

Human impact on tropical forests has changed in pace and scale through time. While about a quarter of a million indigenous people survive in Amazonia today, it is estimated that when Europeans first arrived over 500 years ago, there were perhaps 6 million people living there (Carneiro 1988). Indigenous populations have decreased enormously, largely due to diseases introduced by European settlers. Deforestation and displacement also contributed to decreased populations of indigenous peoples. The Amazonian forest is now less than half the size it was when the first Europeans arrived. Today some tribes persist, carrying on slash-and-burn agriculture and using a variety of plant and animal products from the forest. They cut and burn a small area of forest, generally less than 1 ha in size. They grow crops in the soil enriched by organic matter remaining in or on the soil. After 2–4 years, when weeds invade the site and the organic matter disappears, they abandon it and start the process again.

When population densities are low, and there is no pressure to shorten fallows, slash-and-burn as practiced by these traditional societies has proved to be a long-term system. It has provided the basis for a distinctive culture that, wherever left untrampled by the outside world, continues to flourish. However, these pristine areas are being reduced and threatened with extinction. In some areas of Amazonia, such as the Upper Xingú in central Brazil, a reservation protects the indigenous peoples living there, although a certain degree of acculturation cannot be prevented (Carneiro 1988). However, in most part of Amazonia, indigenous peoples have no such protection.

The Lacandon Maya, still inhabiting forest land in the Selva Lacandona in eastern Chiapas, Mexico, are another example of an indigenous culture whose forested territory has been drastically reduced. Their forest area of about 6,000 km^2 is interspersed with pasture and agricultural clearings (Nations 1988). The central feature of the Lacandon's traditional rain-forest management is a system of agroforestry that produces food crops, trees, and animals on the same plot of land simultaneously. Lacandon agroforestry combines up to 79 varieties of food and fiber crops grown in small garden plots cleared from tropical forest. Like forest farmers throughout the tropics, they create these plots by felling and burning the forest to clean the plot of insects and weeds and to create a temporarily fertile soil by transforming the nutrients of the vegetation into a fertile ash. To conservationists and government planners, the most intriguing aspect of the Lacandon farming system is that they maintain the same plot in production for up to 7 years, while most immigrant colonists have to clear new plots in the forest almost annually. Intensive tending of the plot to keep it clear of weeds is one of the major factors responsible for the long duration of their plots. When the old plots decrease their productivity, the Lacandon plant tree crops and continue to harvest the products while the system remains as a fallow plot and recovers its productive capacity.

Other forest dwellers enrich the primary forest by planting or tending useful species, to create what are sometimes called "agroforests." In southern Sumatra, the forests are enriched with *Shorea javanica* which is tapped for its resin (Whitmore 1998). In Borneo, forests have been enriched with another *Shorea* species for nut production, or with rattan (Brookfield 1993). In the "varzeas" or flooded forests of the delta of the Amazon River in the state of Pará, Brazil, the *Euterpe oleracea* palm or "açai" mentioned above is managed by "caboclos", an amalgam of Indians, African slaves, and European pioneers (Montagnini and Muñiz-Miret 1999).

A number of economic and social factors that are operating today in the major tropical regions of the world threaten the integrity of the rain forest and put in question the continued existence of the traditional lifestyles of forest farmers and hunter-gatherers. Changes that have occurred over the last century have already had a profound effect on traditional ways of life of most forest people. These changes may only accelerate in the near future, resulting in the possible acculturation and subsequent loss of unique human cultures, and knowledge of how to manage the forests without destroying them.

1.5 The Need for an Integrated Approach to Forest Conservation and Management

Tropical forestry is confronted today with the task of finding strategies to alleviate pressure on remaining forests and techniques to enhance forest regeneration and restore abandoned lands. In addition, sustainable forestry in tropical countries must be supported by local and/or international policies adequate to promote and maintain specific activities at local and regional scales. Any efforts to manage, conserve, or restore tropical forests must be compatible with local livelihood needs and offer sustainable and economically attractive alternatives to local people. Strategies that provide various ecosystem services and fulfill local human needs, in addition to promoting the conservation of forest resources, will have a higher chance of success. Understanding the variety of ecosystem products and services of forests is therefore essential for the design of adequate conservation and management strategies, as well as for the development and maintenance of the policies that sustain them.

In conclusion, environmental and social services of tropical forests are as important or more important than activities that degrade the forest structure such as heavy logging or excessive ecotourism. In order to manage the forest to sustain environmental and social services, we must understand the ecology of the tropical forest, in other words, how it works. The understanding of the ecology of tropical forests is essential for designing adequate strategies for forest conservation and management.

Characteristics of Tropical Forests 2

2.1
Characteristics Relevant to Management and Conservation

An objective of this book is to present practices for sustainable management and conservation of tropical forests. Sustainable management means utilizing the forest for human benefit without destroying the capacity of the forest to reproduce itself. Certain ecological characteristics render tropical forests particularly susceptible to disturbances and should be considered if a tropical forest is to be managed sustainably. They are:

- high diversity of species
- high frequency of cross-pollination
- common occurrence of mutualisms
- high rate of energy flow through primary producers, consumers, and decomposers
- a relatively tight nutrient cycle

2.2
High Diversity

The phenomenal diversity of plants and animals in the tropics was first brought to the attention of Europeans by the missionaries and explorers in the 16th and 17th centuries. Eighteenth and 19th century scientists began documenting the biological riches of the tropics and proclaimed them to be extraordinary (Chazdon and Earl of Cranbrook 2002). Although it was not until the 20th century that the concept of species diversity became generally understood, Alfred Russell Wallace in his 1878 book, *Tropical Nature and Other Essays*, recognized high diversity when he recounted his difficulty in finding two individuals of the same species in tropical forests.

> The primeval forests of the equatorial zone are grand and overwhelming by their vastness, and by the display of a force of development and vigour of growth rarely or never witnessed in temperate climates. Among their best distinguishing features are the variety of forms and species which everywhere meet and grow side by side, and the extent to which parasites, epiphytes, and creepers fill up every available station with peculiar modes of life. If the traveler notices a particular species and wishes to find more like it, he may often turn his eyes in vain in every direction. Trees of varied forms, dimensions, and colours are around him, but he rarely sees any one of them repeated. Time after time he goes towards a tree which looks like the one he seeks, but a closer examination proves it to be distinct. He may at length, perhaps, meet with a second specimen half a mile off, or may fail altogether, till on another occasion he stumbles on one by accident.

For almost all types of organisms, the number of species that existed in the tropics was far higher than the number these scientists were accustomed to seeing in their homelands. H.W. Bates (1864) in his book, *The Naturalist on the River Amazonas*, wrote:

> I found about 550 distinct species of butterflies at Ega. Those who know a little of Entomology will be able to form some idea of the riches of the place in this department, when I mention that eighteen species of true *Papilio* (the swallow-tail genus) were found within ten minutes' walk of my house. No fact could speak more plainly of the surpassing exuberance of the vegetation, the varied nature of the land, the perennial warmth and humidity of the climate.

Perhaps the most famous commentator on tropical species diversity was Charles Darwin, who in his 1855 classic *Journal of Researches into the Natural History and Geology of the Countries Visited During the Voyage of H.M.S. Beagle Round the World* wrote as follows about the Galápagos Islands:

> I have not as yet noticed by far the most remarkable feature in the natural history of this archipelago; it is, that the different islands to a considerable extent are inhabited by a different set of beings. My attention was first called to this fact by the Vice-Governor, Mr. Lawson, declaring that the tortoises differed from the different islands, and that he could with certainty tell from which island any one was brought. I did not for some time pay sufficient attention to this statement, and I had already partially mingled together the collections from two of the islands. I never dreamed that islands, about 50 or 60 miles apart, and most of them in sight of each other, formed of precisely the same rocks, placed under a quite similar climate, rising to a nearly equal height, would have been differently tenanted; but we shall soon see that this is the case.

In addition, the many finch species he encountered in the Galápagos Islands, many more than in his native England, stimulated his thinking on evolution.

Tree species diversity in tropical forests is perhaps better documented than diversity of other species, simply because trees are easier to see and count. Some tree species in rain forests are common, but most are rare: in the richest rain forests, every second tree is a different species (Whitmore 1998). It is quite common to find just a few individuals of each species per hectare. For example, a 40-m, or even a 70- to 80-m average distance between trees of the same species may be common. In eastern Sarawak, the density of three tree species averaged 3.6 trees/ha, and 4–5 trees/ha was considered to be high density (Jacobs 1988).

Even the best-represented tree species comprise a low proportion of the total number of species, perhaps a maximum of 15%. Some tropical forests are called by the name of a species that is characteristic of that forest, but is not necessarily dominant. For example, Richards (1996) describes the *Mora* forest in Trinidad and the *Eperua falcata* forest in Surinam, concluding that their presence is due to some limiting environmental conditions, mainly soils of poor drainage. A similar pattern is found in the *Carapa guianensis* forests in the Caribbean lowlands of Costa Rica, that are found on swamps, and with the *Pentaclethra macroloba* forests at La Selva Biological Station in Costa Rica (Hartshorn and Hammel 1994).

2.2.1 Latitudinal Gradients of Species Diversity

Is species diversity really higher in the tropics than at higher latitudes? Most landscapes familiar to European explorers were disturbed by human activity. The diversity in tropical regions might have appeared high to these explorers, simply because tropical forests were less disturbed by human activity than forests in their European homeland, where the goal of forest management frequently was to eliminate "non-economic" species from the forest. To answer the question, scientists began to conduct surveys in relatively undisturbed areas along latitudinal gradients. By the 1980s, it was possible to show that there is, in fact, a gradient of increasing diversity with decreasing latitude. Figure 2.1, based upon the number of plant species encountered in 0.1-ha

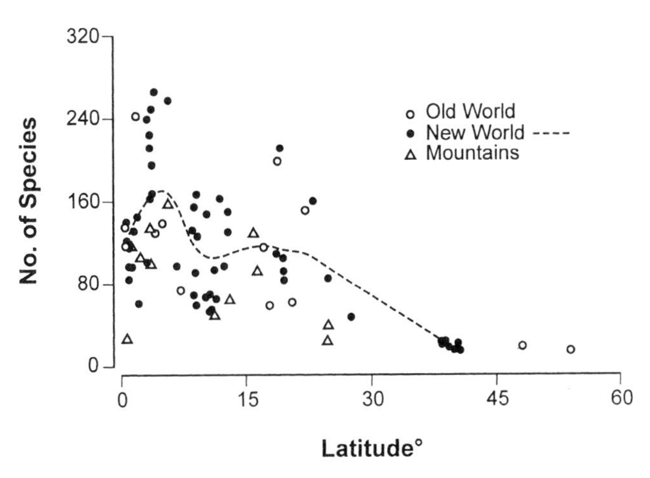

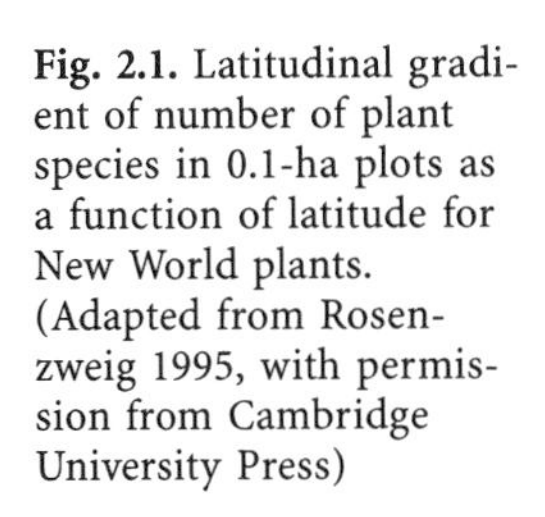
Fig. 2.1. Latitudinal gradient of number of plant species in 0.1-ha plots as a function of latitude for New World plants. (Adapted from Rosenzweig 1995, with permission from Cambridge University Press)

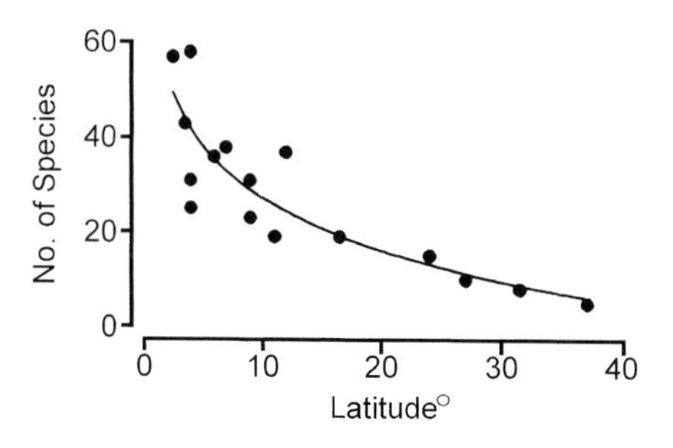

Fig. 2.2. Latitudinal gradient of number or termite species. (Adapted from Rosenzweig 1995, with permission from Cambridge University Press)

plots located at various latitudes, shows that tropical forests are in fact richer in species compared with forests of higher latitudes. Despite significant scatter, the trend toward higher diversity at low latitudes is clear.

Animals as well as plants show the trend of increasing diversity with decreasing latitude. Figure 2.2 illustrates the trend for termites. Similar patterns for bats, mammalian quadrupeds, snakes, frogs, lizards, fishes and fossil Foraminifera are shown in Rosenzweig (1995). Fischer (1960) presents evidence that ants, nesting birds, and a variety of invertebrates are more diverse at low than at high latitudes. Diversity of insect herbivores in the wet tropics is extremely high. Erwin (1982) estimated that 1 ha of forest in Panama may have in excess of 41,000 species of arthropods. In tropical forests, there is a high diversity of organisms that decompose leaves and wood.

2.2.1.1 The Latitude Effect

Rosenzweig (1995) attributes the high scatter in Fig. 2.1 in part to the small sampling areas (0.1 ha). However, other factors such as soils and local climate may be more important for explaining the high scatter of diversity at any given latitude. This raises the question, what is the factor that determines the latitude effect? The number of daylight hours during the year varies very little from equator to poles, and, consequently, the total annual amount of solar energy impinging on the earth's surface also varies very little. However, there is a great difference in the amount of light reaching the Earth's surface per year, when temperatures are above freezing. There is a strong decrease in the amount of light when temperatures are above freezing, along a line from equator to poles. Consequently, there is much less light available for photosynthesis at high latitudes than at low latitudes. There are also lower average yearly temperatures. However, average daytime temperatures during the growing (frost-free) season vary very little along a latitudinal gradient (at locations at the same elevation), so this cannot be the factor that causes differences in diversity. Light available during the growing season is the factor that constitutes the latitude effect (Jordan and Murphy 1978).

Because of the large variations in soils, rainfall, and temperature as affected by elevation that can exist within single latitudes, for an examination of the latitude effect on species diversity, the comparison must be restricted to ecosystems with similar soils, similar rainfall, and at similar elevations. The gradient of increasing diversity with decreasing latitude can only be found when comparing similar types of forest ecosystems such as lowland moist forests on good soils. To show unequivocally that species diversity changes as a function of latitude, all other variables must be held constant (Box 2.1).

Box 2.1

Latitudinal gradients of species diversity

If latitude (yearly light available for photosynthesis) influences diversity, and if effects of soils, rainfall, and temperature as affected by altitude are eliminated, there should be a linear increase in the number of species per unit area as one approaches the tropics. It is quite difficult to find a region where these other factors are constant along a latitudinal gradient. However, Rosenzweig (1995), using data from Specht (1988), was able to show a latitudinal gradient in diversity of plants in sandy coastal habitats in Australia (Fig. 2.3), a single uniform biotope.

Likewise, Lewis (1991) found a gradient of increasing floristic richness with decreasing latitude in the subtropical forests of the eastern Chaco region in Argentina. In these forests, at an intermediate scale, different forest types are arranged according to environmental gradients correlated with topographic elevation. At a fine scale, many micro-sites can be discerned with different microenvironments colonized differentially by species. It is the greater number of species in the micro-sites at lower latitudes that results in the diversity gradient.

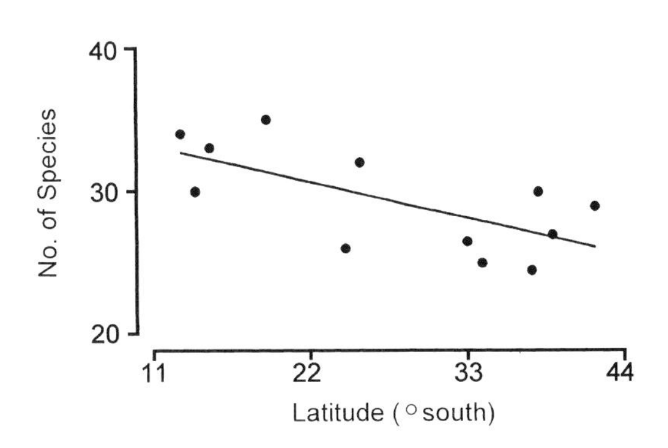

Fig. 2.3. Latitudinal gradient of number of plant species under similar conditions of climate and soils in Australia. (Adapted from Rosenzweig 1995, with permission from Cambridge University Press)

2.2.2 Effects of Elevation on Species Diversity

A variety of terms refer to forests of high elevation in the tropics. Cloud forests include all forests in the humid tropics that are frequently covered in clouds or mist, thus receiving humidity through the capture or condensation of water droplets (horizontal precipitation) which influences the hydrological regime, radiation balance, and several other climatic, edaphic, and ecological parameters (Stadtmüller 1987). Cloud forests may include lower montane (about 1,000–2,000 m), montane (2,000–3,000 m), montane thicket, and finally elfin woodland, until the treeline at about 4,000 m, depending on latitude. With higher elevation, the forest decreases in stature and in number of species. For example, in lowland Malaya (about 400 m maximum elevation) there are nine distinct forest types according to the predominance of one species or another; higher up, there are six forest types, and the next zone up has only two forest types (Jacobs 1988).

In general, cloud forests are found at elevations higher than 1,500 m above sea level; however, on some islands like Puerto Rico and Jamaica and in some isolated mountains (e.g. the Macuira Mountain in Colombia, and in the Santa Ana Mountain in Venezuela), cloud forests can be found at a much lower elevation (Grubb 1977). These elevations may vary depending on the "Massenerhebung effect", the compression and lowering of life zones on small land masses as compared to continents (Grubb 1977), due to a faster saturated lapse rate (moist adiabatic rate). The moist adiabatic lapse rate is the rate at which saturated air cools as it is lifted up by the wind or the air currents. As moist air is lifted up and the temperature decreases, the air becomes saturated and cloud formation occurs. The moist adiabatic lapse rate is lower when air encounters large land masses in part because large land masses radiate more heat than small masses. For example, on Caribbean islands, the lapse rate is high because land masses are small. In contrast, along the west coast of South America, air must rise to a higher level before dew point saturation, due to the large mass of the Andes.

On the island of Puerto Rico, a wide variety of ecosystems exist each with distinct species. On an elevational gradient, species diversity is greatest at 700 m, at the transition between the lower and upper montane forests where cloud condensation is common. Diversity decreases with increasing altitude through palm forest to the cloud forests which begin here at only 1,000 m (Lugo and Scatena 1995; Weaver 1995).

Along an elevational gradient beginning on the west coast of South America and up the Andes, we would expect a series of plant communities as follows (van der Hammen 1974): savanna and dry tropical woodland with low species diversity; lower tropical forest with increased precipitation and increased diversity; sub-Andean and Andean forest, where temperature, moisture, and soils are optimal for plant growth and species diversity is highest; sub-paramo and paramo where temperature limits metabolic rates and lowers decomposition, resulting in decreased nutrient availability, which decreases species diversity; and finally perennial snow where plants are limited to a few species of algae and lichens. Moisture condensation commonly occurs between 2,000 and 3,500 m (Lauer 1993), where cloud forests typically begin. The gradient down the eastern slope is similar, except in the lower reaches where low soil fertility rather than lack of moisture results in lower species diversity (soils in the Amazon basin are highly weathered and low in nutrient elements when compared to younger soils in the Andes). Each forest type along the elevational gradient has its own complement of species, and the nature and the diversity of these species characterize each forest type.

2.2.3 Effects of Soil Fertility on Species Diversity

Soil fertility can vary greatly at any given latitude and can affect species diversity. For example, Stevens et al. (2004) found that in Great Britain, species adapted to infertile conditions were systematically reduced at high levels of nitrogen deposition. In earlier studies in which fertilizers were added to small plots, enrichment resulted in a loss of species diversity (Huston 1979). However, Tilman (1987) pointed out that lower diversity in enriched plots occurs because plant communities do not have time to adapt to changed conditions. When plant communities have evolutionary time to adapt to soil conditions, those on richer soils will have higher diversity than those on poor soils. A decrease in diversity following addition of fertilizer merely indicates that certain species are better able to take advantage of changing conditions than others. The former gain a competitive advantage and crowd out the latter. However, this is a short-term result. When we speak of high diversity on rich soils, we refer to an ecosystem where there has been time for species to immigrate or to adapt. Thus the number of plant species on highly weathered soils of the lowland rain forest of the upper Rio Negro in Venezuela (Clark and Liesner 1989) is lower than those found on younger, richer soils of Ecuador (Gentry 1988).

However, species richness in tropical forests may not always peak in the richest soils. Davis and Richards (1933–1934) found that in wet seasonal lowland Guyana, species richness peaked in mixed forest on moderately low-nutrient yellow sand Ultisols, and was somewhat lower on richer alluvial loams, and much lower on the acid, nutrient-poor Spodosols. Rich soils also will have lower biodiversity at high latitudes than rich soils at the equator, because of the latitude effect. Poor soils, such as those of the Galapagos Islands, often have a high incidence of endemism, but endemism is not to be confused with diversity. An ecosystem can be high in endemics but low in diversity, compared to other ecosystems at the same latitude.

2.2.4
Influence of Stress on Species Diversity

In general, species diversity is lower in ecosystems that are stressed than in those in which conditions are optimal for life. Thus ecosystems in dry regions have lower diversity than those where rainfall equals potential evaporation. For example, in the dry forests on the southwestern coast of Puerto Rico, diversity of trees is lower than in the rain forest of the eastern end of the island (Murphy et al. 1995). However, diversity can be low where excess rain causes soils to be frequently saturated. Nutrient imbalance in soils can also cause stress: ecosystems on nutrient-poor soils have fewer species than those on richer soils. Ecosystems on soils with high levels of potentially toxic elements such as selenium are also relatively low in species. Salt marsh ecosystems have low diversity because of the stress of salt and tides. Low temperatures and large daily fluctuations in temperature at high elevations cause stress, resulting in lower diversity at higher elevation. Highest diversity is almost always found in ecosystems that are not stressed, that is, they are well watered, have deep rich soils, and optimum temperatures for photosynthesis.

Some ecologists may argue that there is no such thing as a natural ecosystem that is stressed because there are always evolutionary adaptations to conditions such as infertile soils, low rainfall, and extreme temperatures, and thus the plants that live under these conditions cannot be classified as stressed. The problem is with the definition of stress. There are environmental conditions where plant growth is best. These optimal conditions consist of water availability that exceeds evapotranspiration but not to the extent where the soils are waterlogged. Nutrient elements are present in sufficient amounts in the soil. Temperatures are optimal, that is, somewhere between 22 and 30 °C, depending on the species (Aber and Melillo 1991). Above the optimal temperature, respiration exceeds photosynthesis and growth decreases. Below the optimal temperature, metabolic rates are low. Thus both high and low temperatures can be considered stressful. For the purposes of discussion, stress can be defined as any condition deviating from the optimum for plant growth. When we take this definition of stress, we can say that diversity generally decreases along a gradient of increasing stress, that is, where climate becomes very hot or very cold, rainfall is low or very high, and where soils are low in certain essential elements, but high in others, species diversity will decrease.

2.2.4.1
Other Factors Influencing Diversity

Another factor influencing diversity is the fact that it is usually lower on islands than on continents, even those of the same latitude. For diversity to be maintained in a particular location, there has to be the possibility of immigration from neighboring locations, because in any location, there is a finite rate of extinction. On island ecosystems, the probability of immigration of a species is low. Once a species does arrive, its probability of extinction is high, because of the difficulty for other individuals of the same species to reach the island (MacArthur and Wilson 1967).

Up to this point, we have been talking about diversity at the species level, that is, diversity of species within a community. Genetic diversity is another component of overall diversity, but determining genetic diversity is much more difficult than determining diversity at the species level. There is yet insufficient evidence to conclude that genetic diversity of trees in the tropics is higher than that of temperate-zone trees (Bawa and Krugman 1991).

2.2.5 Theories to Explain High Diversity in the Tropics

The question, "Why are there so many species in the tropics?" is one that has intrigued scientists for almost a century. Scores of theories have been put forth to account for the increase in diversity in almost all taxa along a gradient of decreasing latitude. Pianka (1966) reviewed the major theories of the time and lumped them into six categories: the time theory; the theory of climatic stability; the theory of spatial heterogeneity; the competition hypothesis; the predation hypothesis; and the productivity hypothesis.

The time theory assumes that all communities tend to diversify in time, and that older communities therefore have more species than younger ones. Temperate regions are considered to be impoverished due to recent glaciations and other disturbances. However, Deevy (1949) argued that only in cases where barriers to dispersal are pronounced can the ecological time theory be of importance in determining species diversity. Where there are no barriers, species can spread rapidly.

A hypothesis that has been popular is that there are more species in the tropics because there are more ecological niches. Fischer (1960), in his review of the concept, explained as follows: "A given environment provides a variety of possible ways for organisms to make a living, and the organisms themselves greatly multiply the number of these *ecologic niches*, in which properly adapted species can prosper and procreate." Because the lowland tropics have been least affected by climatic fluctuations in geological history, there has been more time for species to exploit all the available niches.

The theory of climatic stability, similar to the niche theory, hypothesizes that because of the relative constancy of resources, regions with stable climates allow the evolution of finer specializations and adaptations than do areas with more erratic climatic regimes (Klopfer 1959). This results in "smaller niches" and more species occupying the unit habitat space. However, it is not clear whether climates in the tropics actually are more stable when stability is defined as deviation from an average. A cyclic weather pattern also can be defined as stable if the cycle is regular. Even if stability is defined as deviation from a physical average, stability can be low in the tropics when it is defined as deviation from an average that a species can tolerate.

The theory of spatial heterogeneity assumes that there is a general increase in environmental complexity as one proceeds towards the tropics. The more heterogeneous and complex the physical environment, the more complex and diverse the plant and animal communities supported by that environment tend to be. However, it has been difficult to show that the tropical environment, on the scale in which diversity is usually measured (1 ha or less), is any more complex than the environment at higher latitudes. For a regional scale, it can be argued that there are more habitats in the tropics than at high latitudes (Simpson 1964). For example, Costa Rica has a whole range of habitats from low-altitude tropical to middle-altitude temperate to high-altitude boreal habitats, whereas regions of higher latitude progressively lose some of these habitats. Janzen (1967) explained the reason for this by pointing out that it is the seasonality at high latitudes that causes reduction of habitats on mountains. Species there are better adapted to fluctuating temperatures and can migrate more easily up and down slope. On tropical mountains, species do not have to adapt to seasonal change, and therefore their range is often restricted to a particular narrow band of temperatures that occur at a particular elevation. Temperature barriers along an elevational gradient are therefore greater in the tropics, and mountainsides can be partitioned into more niches, each with its own complement of species.

However, the question of micro-spatial heterogeneity in the tropics vs. higher latitudes remains unanswered. How is micro-spatial heterogeneity defined? If it is defined as the number of species that occupy a space, then the theory becomes circular. There are more species in the tropics because tropical ecosystems have more niches. However, the number of niches in an ecosystem is determined by the number of species in that ecosystem.

The competition hypothesis is based on the idea that competition is the most important factor of evolution in the tropics, whereas natural selection at higher latitudes is controlled mainly by physical factors such as drought and cold (Dobzhansky 1950). Such catastrophic mortality factors are said to be rare in the tropics and thus competition for resources becomes keener and niches become smaller, resulting in a greater opportunity for new species to evolve. However, there is little evidence that catastrophic events, especially droughts, are any less common in the tropics than at higher latitudes.

The predation hypothesis contradicts the competition hypothesis. It claims that there are more predators and/or parasites in the tropics and that these hold down individual prey populations enough to lower the level of competition between and among them. The lowered level of competition then allows the addition and co-existence of new intermediate prey type, which in turn support new predators in the system. However, the predation hypothesis does not explain why there are more predators and/or parasites in the tropics to begin with. If there are more predators because there are more prey species, but there are more prey species because there are more predators, then the argument is circular.

The productivity hypothesis states that greater production results in greater diversity (Connell and Orias 1964). The idea is that in regions where productivity of plant species is high, more food is available for herbivores. Species that would not survive in areas of low productivity can survive in the tropics because there is an excess of available energy. The hypothesis rests on the observation that there is a correlation between high primary productivity and high diversity in many ecosystems. However, correlation does not necessarily prove cause and effect. A third unexamined factor such as rich soil could result in both high productivity and high diversity.

According to Rosenzweig (1995), Terborgh (1973) "cut the Gordian knot" of interwoven explanations for high species diversity in the tropics. The tropics, he noted, are richer than any other place because they are more extensive than any other place. He noted that the land area of the northern and southern tropics is roughly double that of any other zone. With so much more territory to explore, there is much more opportunity to harbor species. However, temperate and tropical regions of comparable size differ in diversity (MacArthur 1969). For example, there are thousands of square miles of land in North America, Europe, and Asia that do not have diversity comparable to the tropical region.

Recently, Hubbell (2001) has criticized the niche assembly rules to explain patterns of biodiversity. He proposed that dispersal assemblies, that is, groups of species that have become assembled purely on the basis of which seeds happened to reach a particular location have equal or greater importance than differences in microhabitat (niche) in the observed pattern of species distribution. The idea is that all species at the same trophic level (for example, trees) are ecologically equivalent and that dispersal of seeds from parent trees can account for observed patterns of diversity. However, Condit et al. (2002) found that the dispersal theory alone cannot account for species distributions except in small uniform areas that have been colonized by early and mid-successional species that do not require particular environmental conditions, as do mature forest species.

Local disturbance has been suggested by Connell (1978) as the explanation for high diversity in tropical forests. His hypothesis, called the "intermediate disturbance hypothesis", postulates that maximum diversity occurs in ecosystems that are subjected to intermediate regimes of disturbance. Molino and Sabatier (2001) tested the hypothesis in French Guiana and found that diversity was higher in areas with light-intensity disturbances such as tree falls and selective cuts. However, to say that these disturbances are the cause of greater diversity in the tropics is to ignore the element of scale (Willis and Whittaker 2002). Most of the species that invade disturbed areas in the American tropics are pioneer species. On small-scale plots (20×20 m) such pioneers will increase the diversity if there has been a disturbance such as a tree fall within these plots. However, these species are extremely common and widespread. Therefore, they will increase the diversity of a plot where disturbance occurs, but they will not increase the number of species occurring at the landscape level.

Intermediate levels of disturbance will not increase the diversity of local endemic species in a mature forest because these species have adapted over the millennia to the conditions of a mature forest. Intermediate disturbance will increase diversity due to immigration of common pioneer species. Diversity can be high in a disturbed area following the invasion of pioneer species in a mature or "climax" community (Whittaker 1975). Conservationists are generally more concerned with preserving species endemic to mature communities. Pioneer species are rarely considered endangered species. Logging, shifting cultivation, and other anthropogenic disturbances ensure the survival of pioneer species, often considered to be weeds.

In the end, there is little agreement on the reasons for high diversity in the tropics. Perhaps the Eurocentric perspective of scientists caused them to look at the problem from the wrong viewpoint. Perhaps more progress could have been made on the diversity question had there been a scientist from the tropics who, while traveling in Europe or North America, would have asked, "Why are there so few species at high latitudes?" Then it would be clear from the evidence in Section 2.1 that diversity tends to decrease with increases in stress. Diversity is highest where stress is least, that is, where temperatures are optimum year round, rainfall is adequate, and nutrients are plentiful and well balanced.

2.2.6 Benefits of High Diversity

2.2.6.1 Defense Against Pests and Diseases

The wet tropics, with their almost continuous hot, moist conditions, are an ideal environment for growth of bacteria and fungi, many of which cause diseases, and the insects that carry these organisms. At high latitudes sub-freezing temperatures can reduce or eliminate populations of disease-causing organisms and herbivorous insects. At the beginning of each growing season, population growth for many insect and disease species must start again. This defense against population explosion is missing in continually moist tropical forests. In tropical areas where there is a distinct dry season, fire may play a role similar to that played by freezing temperatures at high latitudes. For example, in northern India there are pure stands of *Shorea robusta* (sal) due to periodic fires which favor the fire-resistant sal and eliminate competing trees (Goldhammer 1993). Such virtual monocultures are rare or non-existent in the moist tropics.

Because of the high herbivore pressure on many tropical plants, a variety of defense mechanisms have evolved. One that is common in tropical plants is the presence of secondary plant chemicals that make leaves unpalatable to many herbivores (Coley 1980). Another defense mechanism of tropical forests against pests and disease outbreaks is a characteristic of the ecosystem: high species diversity. High diversity of tree species in the tropics provides survival benefit for the forest. Many herbivores and diseases are specific to a particular species. If an insect or disease organism locates and attacks an individual of a certain species, the high diversity of the forest makes it difficult for the pest to locate other individuals of the same species. The greater the distance between individuals of a given species and the greater the number of other species between individuals of the same species, the lower the probability that an insect or disease organism will find the next individual, and the lower the rate of population growth of disease organisms and herbivores (Janzen 1970). That is why pure stands of native species are rare or non-existent in the wet tropics, at least in areas that are not stressed by factors such as soils that are saturated, salty, or extremely poor in available nutrients. Diversity is a naturally occurring defense against disease and insect attack (Box 2.2).

Box 2.2

Diversity as a defense against plant disease and pests

The mahogany shoot borer, *Hypsipyla grandella*, is a particularly noxious pest in forest plantations throughout the American tropics. Larvae of this insect bore into the stems and terminal shoots of young plants, particularly genera such as *Cedrela* (Spanish cedar) and *Swietenia* (mahogany). As a result, seedlings in monoculture plantations can be killed or severely distorted. A number of approaches to control the pest have been tried, including biological control, genetic engineering of host plants, and sustained release of systemic chemicals (Orians et al. 1974; Navarro et al. 2004). However, planting seedlings at wide spacing within native forests perhaps is most effective because it mimics the protection that the trees have under natural conditions. Planting *Cedrela* trees in agroforestry systems with coffee has been shown to be an effective way to avoid insect attack. In experimental research in Costa Rica the attack of the shoot borer insect was lower when *Cedrela* trees were planted in adult coffee plantations rather than in recently planted or pruned coffee (Navarro et al. 2004). Mixed plantations and agroforestry systems are discussed in more detail in Chapter 6.

There could be objections to the idea that the high plant diversity in moist tropical forests is an adaptation to herbivore pressure. Adaptation takes place at the species level and diversity is an adaptation at the community level. However, adaptation here is not used in the classical sense. It simply means that low-diversity forests are rare in the wet, fertile tropics because it is difficult for them to become established. The proximity of individuals of the same species makes them easy prey for outbreaks of disease. Diverse communities are less susceptible to sustained attack and destruction by pests and diseases.

The diversity of species in tropical forests not only helps to ensure the survival of each species, it also helps to ensure the continued existence of tropical forest ecosystems. While individual populations within a forest peak and crash (May 1972), the flow of energy and cycling of nutrients within a forest remain relatively stable. At each trophic level, there is a continual dynamic flux, as the population of one species that is rising replaces the population of another that is falling (McCann 2000; Tilman 2000).

2.2.6.2 Complementarity

When several species of plants are growing together and the yield per hectare is greater than that of a pure stand, the phenomenon is called overyielding. Overyielding may result from more efficient use of resources – land, nutrients, water, and sunlight – because the various species complement each other in the way they use these resources (Trenbath 1986). Some species have roots concentrated near the surface, while others have deep roots. When they grow close to each other, one can take up nutrients and water unavailable to the other. Different species often have different nutrient requirements; one species may need high levels of nitrogen, while another may be a calcium accumulator. Different species have different water requirements. Forests that have a mixture of species with different water requirements may not produce as much biomass as a monoculture of one water-demanding species in a wet year. However, in the long run, the combination of species of low and high demand for water ensures that there will always be at least *some* growth.

Each species has its own characteristic leaf shapes, colors, and angles. As a result, some species are more efficient in the overstory, while some are more efficient in the understory. Some do better with diffuse light, as occurs on a cloudy day, than in direct sunlight. Each species has a distinct and separate niche. When overyielding is caused by a more efficient use of resources an ecosystem is said to have ecological complementarity. We could also say that overyielding in diverse forests results from a lessening of the intraspecies competition that occurs in a monoculture where adjacent individuals all compete for the same resource at the same time. The idea of overyielding and complementarity was first applied to agricultural systems, where it has been shown that a mixture of crops can have a greater yield than a monoculture (Vandermeer 1990). Because of the difficulty of carrying out a test with replicated plots, it is difficult to confirm the idea that overyielding occurs in natural forests (Ashton 2000). However, Naeem et al. (1994) rigorously tested the theoretical advantage of species mixtures and found that high diversity enhances ecosystem functions such as productivity and resistance to disturbance.

Studies of plantation forests in the USA, Europe, and the tropics have shown that once established, mixtures of compatible tree species have higher yields than single-species plantations (Ashton and Ducey 2000), thus lending credibility to the idea that overyielding and complementarity are important in natural forests. Observations of the characteristics of trees that become established in tree-fall gaps of natural forests also suggest differences in the roles that various species play in natural forest dynamics. Tree-fall gaps occur when mature trees fall over due to death or accident and leave an opening in the canopy that exposes the forest floor to sunlight. If the trunk of the dead tree breaks off but the roots remain in the soil, the seedlings and saplings already present will accelerate their growth due to the new availability of light, nutrients, and soil moisture. These trees have been termed strugglers (Oldeman and van Dijk 1991) because they are able to live for many years struggling to survive under conditions of low light and high competition. When the falling tree is uprooted, and a patch of mineral soil is exposed, the pioneer or "r"-adapted trees that colonize have very different characteristics from those of mature forest species. The pioneer-type species, due to their rapid growth, are better able to take advantage of the opening in the canopy than slower-growing mature forest species. These differences reflect different ways in which species capture light, suggesting that species have evolved separate niches to more fully utilize the available light – in other words, there is complementarity of species.

Diversity in time can be as important as diversity in space, especially when early and late successional species complement each other's roles. Many forest species of mature communities are adapted to the partial shade and humid conditions of the undisturbed forest. When these species are planted in the open sun and bare soil, they often do poorly. Young seedlings and saplings of many rain forest trees do much better when they establish under a nurse species that lessens the severity of the microclimate around the seedlings, par-

Fig. 2.4. Complementarity of structure of species in uneven-aged, mixed-species forest

tially suppresses weeds, and provides leaf litter that increases the soil organic matter (Mesquita 1995). Good nurse species often will be early successional species that can establish themselves in harsh microclimate and disturbed soils. Their ecological role complements that of mature forest species.

The mixture of canopy physiognomies illustrated in Fig. 2.4 results from an uneven-aged forest, that is, all the trees did not begin life at the same time. The differences in ages between individuals of different species helps to accentuate the niche differences, because not only are there differences horizontally, there are also differences vertically. In even-aged stands, where trees started growing at the same time, much of the vertical differentiation would be lacking.

2.2.7 Implications of High Diversity for Forest Management

Present rates of species extinction throughout the world are at least several hundred times greater than the rate expected on the basis of the geological record (Dirzo and Raven 2003). The greatest losses are occurring in the tropics. Some of the losses result from invasion by exotic species. For example, the yellow crazy ant, *Anoplolepis gracilipes*, was introduced to Christmas Island in the Indian Ocean about 70 years ago. Populations exploded in the 1980s, forming super-colonies that infested one-fifth of the island. The ants eliminated the red land crab, a keystone consumer in the forest floor ecosystem (O'Dowd et al. 2003). Invasive plants also can cause extinction of endemic species through allelopathy, or the secretion of toxic chemical into the environment (Bais et al. 2003). However, habitat destruction is the most important driver of species extinction (Dirzo and Raven 2003).

Conservation of species is difficult to incorporate in forest management plans. It is not simple to design management strategies when many species have to be considered. For example, when deciding what trees to mark for selective cutting and what trees to leave behind as seed sources for the next generations, managers should know the ecology and reproductive biology of the species they are removing. Usually such information is not available (Bawa and Krugman 1991). Forest management is discussed in detail in Chapter 5.

The situation in the Amazon region illustrates the problem with saving a valuable tree species when individuals of that species are widely scattered in a high-diversity forest. Many tree species in Amazonian forests will not bring in enough money for loggers to bother with them. Mahogany (*Swietenia macrophylla*), however, brings such an extraordinary price that loggers find it worthwhile to bulldoze through miles of low-value hardwoods to locate one mahogany (Castellanet and Jordan 2002). The opening of logging roads or even just skid trails can be extremely destructive and disturbs far more than just the trail and the tree cut. The roads make it easy for land-hungry peasants, ranchers, and land speculators to enter the region. Once the land is cleared, repeated fires and other disturbances make recovery of a forest community extremely unlikely. Even if a secondary forest does become established, many species of primary forest will be absent, because there are few or no parent trees or animals to disperse seeds (Nepstad et al. 1990).

However, there are other ways to take economic advantage of forests so as to benefit the local population while at the same time preserving rare and valuable species. For example, Chapter 5 discusses more technical forestry approaches, such as reduced-impact logging, and in Chapter 6, we discuss community forestry.

2.3 Reproductive Ecology of Tropical Trees

Certain characteristics of the reproductive ecology of tropical trees have important implications for conservation and management. These characteristics include: timing/frequency of flowering and seed production; predominance of cross-pollination; and a high proportion of dioecism (male and female flowers on separate individuals).

2.3.1 Timing/Frequency of Flowering and Seed Production

It is important to conduct phenological studies, i.e. to know when during the year the trees start flowering, fruiting, and producing seed, so that one can collect seeds of desired species that will be later planted in enrichment plantings or in plantations or for agroforestry. In tropical forests, flowering, fruiting, and seed production are most frequent in the dry season. Generally, each species flowers at a certain interval, with regularity. Continuous flowering after the juvenile period is over is found in some secondary forest tree species, for example, *Trema orientalis*, and also in some mangroves, such as *Rizhophora mangle*. However, periodic flowering is more common (Longman and Jenik 1987). Frequency of flowering periods may be 3–4 months to 10–15 years. Many common forestry species flower once a year (for example, *Cedrela odorata*, *Gmelina arborea*, *Tectona grandis*, *Terminalia ivorensis*). Many canopy trees in the rain forest of Surinam have a biennial flowering habit (Longman and Jenik 1987).

Many species of dipterocarps, important timber species in SE Asia, flower every 3–8 years, a pattern that is called "mast" flowering (general or gregarious flowering) and fruiting. In those years, the majority of dipterocarps of the whole region flower profusely and gregariously over a period of a few months, often accompanied by flowering of species of other families as well. Mast flowering also occurs in plantation trees of these species. The environmental trigger for mast flowering is thought to be an increase in the diurnal fluctuation of temperature associated with drier spells of weather with clearer skies (Yap and Chan 1990). Although the pattern is not as marked as in the dipterocarps, several species of the Vochysiaceae family in the Neotropics have an irregular flowering pattern, which is affected by variations in annual rainfall: there may be no flowering at all in some years, some trees may flower at different times during the year, and, in some cases, trees may flower but will not produce viable seeds (Flores 1993). Several species of Vochysiaceae are important timber species; therefore this pattern also influences the success of management in enrichment planting or in plantations utilizing these species.

2.3.2 Modes of Reproduction of Tropical Trees

Tropical trees can be *hermaphroditic*, when both sexes are in the same flower: examples include leguminous species and citrus; *monoecious*, when both sexes are in the same plant but in different flowers, i.e., there are separate male and female flowers on the same individual: examples include pines, palms, and oaks; or *dioecious*, when sexes are on different individuals, i.e., male flowers occur on different trees to female flowers: examples include species belonging to the families Araucariaceae, Myristicaceae, Piperaceae, and Euphorbiaceae, among others. About 60–65% of trees in lowland rainforests have been found to be hermaphroditic; 11–14% monoecious, and 23–26% dioecious (Bawa 1992).

The mode of reproduction has important implications for management and conservation. Research has shown that most tropical woody plants are strongly outcrossed (Bawa 1992). This means that pollen from the flower of one tree fertilizes the flowers of another tree of the same species. In most tropical forests there is a large distance among individuals of the same species. As a result of timber extraction, distance among individuals of the same species may become even larger, and it may be harder for the pollinators to transfer the pollen to another tree of the same species. This poses a limit to the number of individuals that can be extracted from the forest, and also on the minimum size needed for a forest reserve, so as to ensure successful reproduction of the species.

Research on genetic characteristics of tropical trees has shown that there is considerable genetic variation within populations (Bawa 1992). Genetic heterogeneity can be an advantage for withstanding stress and for adapting to new environmental conditions, such as those encountered by trees when they are planted in a plantation. Cross pollination increases genetic heterogeneity. Distance between two individuals of the same species is the main factor affecting cross-pollination. As cross-pollination has great survival value, dioecism is an advantage because it impedes self-pollination. The frequency of dioecism needs to be taken into account when planning the minimum size of area for preservation, since dioecious plants require larger areas than monoecious or hermaphroditic trees. Dioecious trees generally require larger areas, because the chances that the nearest neighbor of the same species is of the opposite sex is 50%, while in the case of hermaphroditic or monoecious trees, the chances that the nearest neighbor will have a flower or flower parts of the opposite sex is 100%. In addition, in the case of monoecious or hermaphroditic trees, if other individuals of the same species are not close enough, the trees may still be able to reproduce by self pollination, although this might result in a decrease in genetic heterogeneity for the next generation. In the case of the dioecious trees, they must necessarily have another individual of the opposite sex close enough to reproduce.

2.4 Species Interactions in the Tropics

Interactions between individuals, species, or functional groups in ecosystems have been classified in various ways. One of the oldest terms for a cooperative interaction between two species is symbiosis. Another term for interactions that benefit the interacting species is mutualism. Boucher (1985) has classified mutualisms into several categories:

1. *Nutrition/digestion*: increasing the availability of nutrients, energy, and water to one or both of the interacting species. An example is mycorrhizae, an association between plants and a fungus through which the fungus obtains carbon for energy, and the plants gain an increased ability to take up water and nutrients from the soil.

2. *Protection* for one or both interacting species. Insects that live on or in a plant and derive energy from the plant may protect the plant in which they live. When an animal begins to browse a protected plant, the insects attack the intruder.
3. *Pollination* of some plants depends upon the actions of specialized insects and birds. The latter benefit from the mutualism by obtaining energy from the plant's nectar.
4. *Seed dispersal*: the spreading of seeds away from the parent tree by birds, bats, and small mammals that eat fruits increases the probability that some seeds will survive and grow.

An important nutrition/digestion mutualism in the tropics is the interaction between termites and fungi. Some species of termite carry material back to their nests where it is decomposed by fungus. The fungus is then used as a food source by the termite colony (Wood 1978). Leaf cutter ants also cultivate similar fungal gardens on leaf fragments (Stradling 1978). The mycorrhiza/tree-root association is also extremely important in the tropics, as it contributes to high efficiency of nutrient cycling critical for tropical forests where there is a high potential for nutrient loss through leaching (Jordan 1985).

Plant–animal mutualisms in which the animals provide protection for the plant are one of the most easily observed mutualisms in the tropics. In some cases an ant colony lives in some hollow or hollowable part of the plant and is directly or indirectly fed by the plant (Janzen 1985). The ants defend the plant against mammalian herbivores by swarming over and stinging the herbivores. As many as 90% of the trees in the Peruvian Amazon have some relationship with protective ants (Marquis and Dirzo 2002). Certain rattan palms have a swollen wooden ligule which houses defender ants. When a large herbivore brushes against the palm, the ants rush out and beat their mandibles on the hollowed ligule. In West Africa, ants inhabiting trees of the genus *Barteria* are capable of stinging an elephant and can numb smaller mammals for several days (Whitmore 1990).

The importance of mutualistic interactions for seed dispersal and pollination appears to increase from temperate to tropical climates. For example, there are no nectarivorous or frugivorous bats north of 33°. Extrafloral nectar glands on plants drop off drastically between the northern limits of the neotropics in northeastern Mexico and Texas. The characteristics of the physical environment may play a role in this pattern. At higher latitudes, where there is a much higher average wind velocity, many plants are wind-pollinated and have wind-dispersed seeds. Trees at high latitudes often do not allocate any energy to mutualistic interactions with animals, since animal vectors are not involved at any stage in their life history. In contrast, in tropical rain forests where wind-pollinated species are rare, many species of animals are supported by pollen, nectar, and fruits (Orians et al. 1974). Dispersal of seeds away from the parent tree is an important function that helps to ensure germination of the seeds and survival of the species (Janzen and Vázquez-Yanes 1991). Seeds that are concentrated close to the parent tree may be more likely to be eaten by herbivores (Janzen 1970). Dispersal by mammals, ungulates, bats, birds, and even fish is common in tropical forests (van der Pijl 1972; Whitmore 1990).

Many of these mutualisms have arisen through co-evolution. For example, in the case of pollination, a newly arisen species of tree might have the ability to be pollinated by a variety of insects or birds. However, one particular species of bird might have proven more successful in pollination than another species, and so, through selection, the tree modified its flower so that it would accommodate only the beak of that particular species of bird. The bird species, in turn, discovered that this particular species of tree offered more nectar than other species and began to restrict its visits to this one species. Through the passage of evolutionary time, the mutualism became obligatory. Neither the tree nor the bird can survive without each other.

Sometimes the interaction is more complicated. A tree may depend upon one species of bird or insect for pollination and on another species for seed dispersion. An insect may depend upon one species of plant as a host for its larval stage and on another for its mature stage (Gilbert 1980). Often, the result of this process may be more diffuse than a simple one-to-one relationship, that is, a species may be dependent on more than one other species for its survival (Whitmore 1990). The complicated food-web mutualisms in tropical forests render the species particularly susceptible to local extinction following disturbances such as logging, even if logging is carefully carried out (Box 2.3).

Box 2.3

Impacts of logging on plant–animal interactions

Logging may inadvertently eliminate individuals of a species upon which pollinators or seed dispersers depend for their various life functions. Pollinators and seed dispersers often depend on a number of specific species for their survival. If only one of these species is eliminated from the range of a pollinator or seed disperser, that mutualist may become locally extinct. If the obligate insect or bird becomes extinct, then the tree species that depends upon it has no chance of reproducing in the wild. However, mutualisms may be less susceptible to extinction when there are spatially differentiated patches that allow a reorganization of the system following a disruption (Bronstein et al. 2003).

Size of openings in the forest, as well as the patchiness of mutualist organization, are important in determining the impact of forest clearing.

When openings in the forest are small, such as in a shifting cultivation plot, re-establishment of mature-forest species can occur as fruits fall into the opening or as seed dispersers traverse it. However, when large areas are cleared, small mammals, birds, and bats hesitate to venture far from the edge of the mature forest (Stouffer and Borges 2001). The loss of seed disperser activity can inhibit the re-establishment of mature-forest species, as has been shown in pastures in Amazonia (Nepstad et al. 1990) and Costa Rica (Holl 1999). In contrast, temperate forest regeneration may be less dependent upon mutualists. Tropical forests tend to be more sensitive to large clear-cuts than temperate zone forests, and logging strategies used in northern forests are therefore not appropriate for the tropics.

Even if the seed-dispersing species is merely reduced in number, reproduction of the tree species may be endangered. The time interval during which seeds of many rain forest species remain viable is often very short (Gómez-Pompa et al. 1972). The fruit or seed must be found soon after ripening. If not, it will germinate or decompose before being transported into the area where reforestation is needed. Other factors, particularly predation, also contribute to low seed germination in disturbed areas. In an Amazon rain forest, seeds of wild species planted in continuous forest were three to seven times more likely to germinate than those in forest fragments (Bruna 1999).

In this section, we have mentioned only a few of the studies that illustrate the importance of mutualisms in maintaining the structure and function of forests through nutrition, protection, pollination, and seed dispersal. There are hundreds if not thousands more. Guides to some of the classical papers in this field are presented by Loiselle and Dirzo (2002) and Marquis and Dirzo (2002).

2.5 Energy Flow

2.5.1 Delineation of the Tropics

The tropics comprise that area of the world between the Tropic of Cancer (latitude 23.5 °N) and the Tropic of Capricorn (latitude 23.5 °S). These parallels mark the points that the sun reaches at its greatest declination north or south. Outside these latitudes, the annual amount of solar energy reaching the earth's surface generally decreases as a function of increasing latitude (Kondratyev 1969). From an ecological perspective, it is more useful to delineate the tropics on the basis of energy balance, as indicated by temperature, rather than latitude. One approach has been to use the mean annual 20 °C

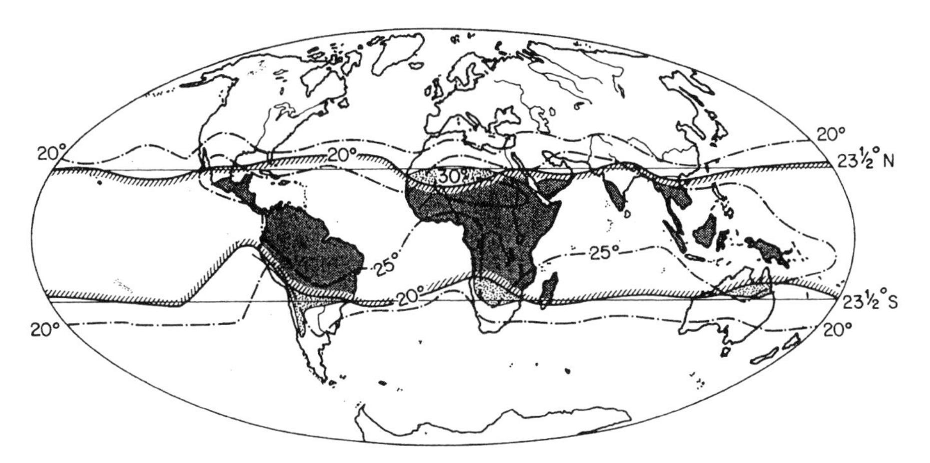

Fig. 2.5. Thermal delimitation of the tropics. The mean annual isotherms of 20, 25, and 30 °C are shown by *lines of alternating dashes* and *dots*. The 20 °C January (northern hemisphere) and July (southern hemisphere) isotherms, corresponding to the coldest months, are shown by *solid lines with hachures*. Areas where the mean daily temperature range exceeds the mean annual temperature range are indicated by *dark shading* inside the 20 °C isotherm and *dots* outside the 20 °C isotherm. (Adapted from Jordan 1985, with permission of John Wiley and Sons Ltd., publisher)

isotherm (Fig. 2.5). Other definitions are: areas with temperatures that exceed 20 °C during the coldest months of the year, and areas that have a mean daily temperature range greater than the mean annual range (Tricart 1972).

2.5.2 Primary Production

In tropical regions where temperatures are continuously warm and rainfall is abundant throughout the year, the growing season is continuous (Jordan 1971a). At higher latitudes, the growing season is shorter. Even though there is impinging solar radiation at higher latitudes or higher elevations, plants cannot use it when daily temperatures drop below freezing. In the dry or seasonally dry tropics, the growing season is also shorter. Lack of moisture inhibits photosynthesis and so the impinging radiation cannot be used by plants.

Because of the year-round availability of moisture and solar radiation, the wet tropics have often been thought of as an environment where forest production would be a maximum. For example, Wallace (1878, p. 65) believed that tropical forests had great productive potential when he wrote, "The primeval forests of the equatorial zone are grand and overwhelming by their vastness and by the display of a force of development and vigour of growth rarely or never witnessed in temperate climates." Ecological studies in the 20th century have confirmed this impression. Net primary production (wood increment plus leaf litterfall) is higher in wet tropical ecosystems than in any other terrestrial ecosystem (Table 2.1). These estimates, however, may be somewhat low. Nemani et al. (2003) found that global climatic changes between 1982 and 1999 resulted in an increase in net primary production in ecosystems throughout the world by 6%. The largest increase was in tropical ecosystems. Amazon rain forests accounted for 42% of the global increase, mainly owing to decreased cloud cover and the resulting increase in solar radiation. In tropical plantations, wood production can be higher than that of native forests (Wadsworth 1997; Dabas and Bhatia 1996). Plantations, however, receive anthropogenic subsidies that can increase their productive potential over naturally occurring forests.

The high rate of net primary production in native tropical forests (Table 2.1) has important implications for nutrient dynamics in the soil, which in turn affects the ability of the soil to sustain production when the forest is logged or converted to plantations or agriculture. High rates of energy transfer to the soil due to leaf litter input and tree fall are accompanied by high rates of decomposition, which in turn results in high rates of carbonic acid formation (see Sect. 2.6.3). This causes relatively high rates of mineral weathering and nutrient leaching, which can decrease the productive capacity of the soil.

The distribution of energy between wood and leaves may be important with regard to energy input to the soil. Leaf litter decomposes more rapidly than wood biomass, due to its greater ratio of surface area to volume and

Table 2.1. Net primary productivity in ecosystems of the world. (Reprinted from Whittaker and Likens 1975)

Ecosystem	Net primary productivity (dry matter)	
	Normal range ($g\ m^{-2}\ year^{-1}$)	Mean (of all studies)
Tropical rain forest	1,000–3,500	2,200
Tropical seasonal forest	1,000–2,500	1,600
Temperate forest		
Evergreen	600–2,500	1,300
Deciduous	600–2,500	1,200
Boreal forest	400–2,000	800
Woodland and shrubland	250–1,200	700
Savanna	200–2,000	900
Temperate grassland	200–1,500	600
Tundra and alpine	10–400	140
Desert and semidesert scrub	10–250	90
Extreme desert – rock, sand, ice	0–10	3

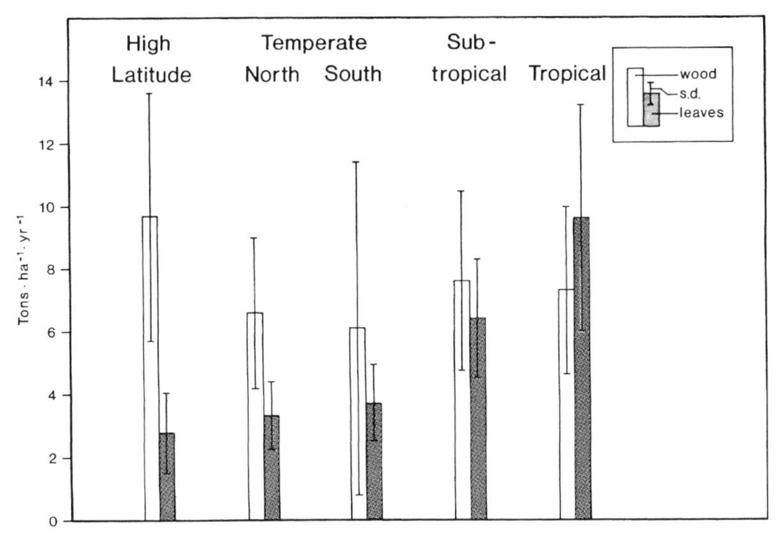

Fig. 2.6. Averages and standard deviations of wood and leaf production for mesic hardwood forest ecosystems at various latitudes. Averages are based on 60 studies of wood production and 135 studies of leaf litter production. Individual data values and references are listed in Jordan (1983). (Adapted from Jordan 1985, with permission of John Wiley and Sons Ltd., publisher)

higher nutrient content. In many tropical forests, leaves comprise a relatively high proportion of primary production (Fig. 2.6). In the Luquillo forest of Puerto Rico, Weaver (1995) found that litterfall comprised 90% of the total net primary production. High rates of leaf input to the soil relative to wood input accentuate even further the high decomposer activity and consequent nutrient transformations compared to forests where the proportion of leaves to wood is lower.

2.5.2.1 Production Patterns Within the Tropics

Geographical patterns of primary production are similar to those of species diversity. Just as diversity is highest where stress is least, production also is highest where optimal conditions of light, nutrients, and water exist. As has been pointed out, optimal conditions exist when temperatures are between 22 and 30 °C, soils are moist but not saturated, and nutrients are present in the soil in sufficient and balanced ratios. In general, productivity declines as elevation increases (Lonsdale 1988). At high elevations, especially in cloud forests, productivity could be limited by saturated soils that limit root respiration, rocky soils, low available solar radiation due to clouds, low availability of nutrients due to their immobilization in thick layers of organic matter, heavy winds,

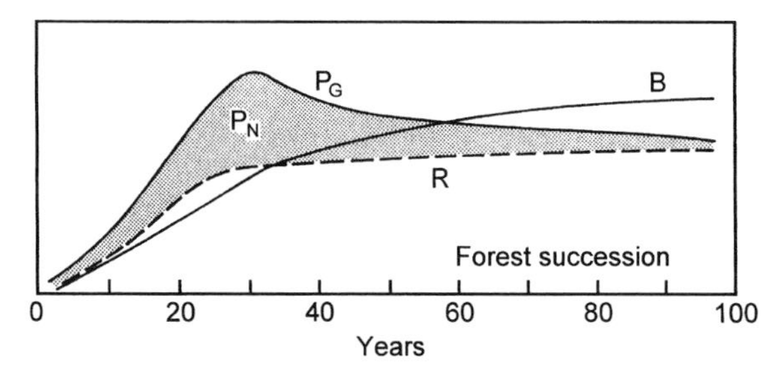

Fig. 2.7. Trends of gross production (P_G), respiration (R), net production ($P_N = P_G - R$), and biomass accumulation (B) in a forest as a function of time during secondary succession. (Redrawn from Kira and Shidei 1967, with permission of the Ecological Society of Japan)

and high concentrations of phenolic compounds in the leaves (Bruijnzeel and Veneklaas 1998). However, on some tropical mountains, primary production can be highest at mid-elevations, if production at lower elevations is limited by high temperatures, nutrient deficiencies in the soil, or a high ratio of evapotranspiration to precipitation (Webb et al. 1983). Nutrient availability also influences primary production. Forests on richer soils produce wood and leaves at almost twice the rate of those on nutrient-poor soils (Jordan 1985).

Productivity also tends to change with time in successional forests. At first, productivity is low, due to small biomass. As the structure of the forest increases, productivity increases, but eventually it declines as costs of maintenance increase (Fig. 2.7). The costs of maintenance include a wide range of defenses against competition, disease, and herbivory, as well as repair of tissues damaged by these factors, wind, lightning, and other physical factors.

Productivity and the accumulation of biomass that occurs depend upon the intensity, size, and duration of the disturbance that initiates the succession (Jordan 1985):

- If the disturbance is of high intensity, large size, and short duration, as would occur on a landslide of a tropical mountain, or after a strong volcanic eruption, then primary succession will occur. It may take decades or even centuries for soil to develop in which even the pioneer tree species can become established. Algae and possibly some grasses will colonize the area. Tree ferns may occupy a stage in humid, montane areas.
- If the disturbance is a hurricane, then the disturbance could be of low to high intensity, possibly large in size, and of short duration. Mature trees that are damaged by hurricanes often have a capacity to resprout. Secondary stages of succession may be entirely bypassed, except in forest gaps where mature trees have been uprooted.

- If the disturbance is of moderate intensity, small size, and short duration, as would occur from a few years of shifting cultivation, then early and late secondary species as well as mature (climax) species may immediately become established. Due to the small size of the forest opening, seeds from the surrounding forest fall into the gap and become established. Large-scale agricultural activities such as banana plantations are much slower to recover once abandoned, because of long distances to seed trees and lack of seed dispersers.
- If the disturbance is one of light intensity, of variable size, and short duration, as would occur with shelterwood forestry (see Chap. 5), then the secondary successional stages would be mostly bypassed. Selective logging as in the shelterwood or similar systems (Fox 1976; de Graaf and Poels 1990) mimics the impact of natural tree fall, with the exception of soil disturbance along the skid trail (path to remove the log).
- If the disturbance is of moderate intensity, large size, and duration long enough to eliminate all sprouting trees from the original forest, then we would expect colonization by pioneer species. Such a disturbance might result from extended periods of agriculture, followed by abandonment with no further disturbance.
- If the disturbance is of moderate intensity, large size, and long duration, as would occur with conversion of forest to pasture followed by periodic burning, then there will be a fire "disclimax". For example, in the Gran Pajonal of Peru, degraded and burned land is first invaded by the fern *Pteridium aquilinum*. If fire continues, the aggressive grass *Imperata brasiliensis* invades. As soil erosion increases, grasses of the genus *Andropogon* become dominant (Scott 1978). Regeneration of forest can be slow, due to lack of seed rain, low seed germination, unfavorable microclimate, and compact and/or nutrient-poor soil (Holl 1999). Recovery to the original forest type may never occur. For example, it is possible that the huge babaçu palm (*Orbignya phalerata*) forests in the Amazonian state of Maranhão (Dubois 1990) represent a "fire disclimax", that is, a community of lesser biomass and species diversity than the original forest in that area. In areas of Southeast Asia, an aggressive grass, *Imperata* spp., often establishes, and is favored by fire (Lal 1987). Once a dense colony of this grass establishes at a site, recovery to forest may be slow or impossible.

Uhl et al. (1990) illustrated the rate of biomass recovery as a function of disturbance type (Fig. 2.8).

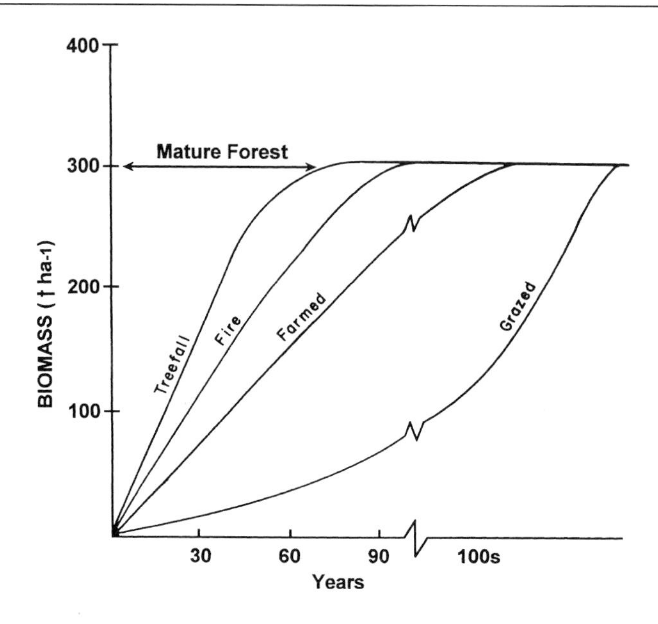

Fig. 2.8. Accumulation of biomass during succession as a function of disturbances of varying intensity. (Adapted from Uhl et al. 1990, with permission of Columbia University Press)

2.5.3 Light Environment of Tropical Forests

Light is the aboveground environmental factor that affects plants the most, and it is also the most variable. For example, on a sunny day, instantaneous measurements of photosynthetically active radiation range over three orders of magnitude, from less than 10 µmol m^{-2} s^{-1} in closed-canopy understory of mature forest to well over 1,000 µmol m^{-2} s^{-1} in exposed micro-sites of gaps and large clearings, or at the top of the forest canopy (Chazdon and Fetcher 1984). The light environment is important for understanding natural forest regeneration and designing techniques for natural forest management. Pertinent questions are: what are the light requirements of the preferred species? How much should the canopy of a forest be opened to allow for increased light in the understory to favor growth of desired species? Knowledge of the light environment is also important for the proper design and management of agroforestry systems and plantations. What species should be in the upper, mid, or lower layers of a multi-strata agroforestry system based on their light requirements? One can learn how to manage the light environment depending on species requirements, or alternatively, how to design a system with species that match the light availability offered by the different microenvironments that are already present in the system.

Attention to the light requirements of species is basic for designing sustainable forest management techniques. For example, it is well known that mahogany is a light-demanding pioneer species that regenerates after catastrophic disturbances such as fires, floods, and landslides that open up large areas of the forest. Most other tree species associated with mahogany have little or no commercial value; therefore, mahogany logging is highly selective and opens up very little of the forest canopy (Lugo et al. 2002). The result is insufficient light levels at the forest floor, a condition inhospitable to regeneration of mahogany. Enrichment planting of mahogany in the understory of natural forests (described in Chap. 6) has had poor results because mahogany seedlings require full light to grow rapidly, a condition found only in larger canopy openings. In most cases, mahogany has responded favorably to silvicultural treatments intended to release mahogany seedlings from competition for light by cutting climbers, girdling surrounding trees, thinning saplings, and pruning.

2.5.3.1 Availability of Light

The portion of radiant energy important for plants is called photosynthetically active radiation (PAR): 400–700 nm (nanometers), which is short-wave, energy-rich radiation. The amount of PAR that reaches a certain area where plants grow is measured in irradiance units, or photosynthetic photon flux density (PPFD), and is expressed in micromoles per square meter per second (Longman and Jenik 1987). Light availability can be expensive to measure in a direct manner, because it requires equipment such as light sensors. Indirect measurements of light availability or total illuminance can be more practical. Indirect measurements of light availability in a forest can include indices of canopy closure based on direct observation of canopy characteristics, or on the use of spherical densiometers consisting of a concave or convex mirror that reflects the canopy of trees and allows estimations of area covered. Hemispheric photography uses fish-eye lenses to obtain pictures of the canopy, and can be complemented with computer software that can calculate the size of the open and canopy areas in a more precise manner. Simple light sensors can also be used for instantaneous measurements, but their reliability is low due to the high variations that can be found in the light environment according to time of day and weather conditions.

Solar radiation is spectrally altered by passage through the forest canopy, affecting both the quantity and the quality of PAR in the understory (Chazdon et al. 1996). Chlorophyll in green plants absorbs some light energy in the blue (400–500 nm) but mainly in the red (600–700 nm) portions of the light spectrum. Leaves (chlorophyll) reflect light around 550 nm (the green band). Leaves are very transparent to far-red (FR, beyond 700 nm) wavelengths, i.e., they reflect and transmit most of the FR radiation that reaches them; thus a high proportion of FR radiation penetrates all the way down to the forest floor. The greatest spectral change after canopy filtering, therefore, is the reduction in R/FR ratio (Chazdon et al. 1996).

2.5.3.2
Responses of Plants to Light

The response of plants to light, i.e., their capture of light energy for photosynthesis, is influenced by several leaf characteristics, including (1) sclerophyllous character (sclerophyllous leaves are more common in forests of nutrient-poor soils; because of their thickness, there is less light absorbed per unit of chlorophyll); (2) predominant leaf size (larger, horizontal leaves are more effective in capturing light, while smaller, vertical leaves on the tops of many tropical trees are more efficient in catching the light from the rising and setting sun); (3) presence of epiphylls (mosses, lichens, and algae) which can interfere in the capture of light; and (4) presence of a layer of anthocyanins (red pigments) which enhance the capture of PAR (Mooney et al. 1984).

Leaf life span is often correlated with overall plant productivity. Long-lived leaves often occur on plants in soils low in moisture or nutrients. In such environments, it is energetically expensive to synthesize new leaves. In lowland tropical forests, leaves often live 3–13 months, while in upper montane forests the leaf life span can be 14–18 months, possibly because in montane forests, limited light or nutrients may inhibit leaf synthesis (Mooney et al. 1984; Jordan 1985). Reich et al. (2004) found that variation in light availability had consistent effects on leaf life span for all species studied in the rain forest at San Carlos de Rio Negro in the Venezuelan Amazon: species native to tierra firme forest in deeply shaded understory microsites, in small gaps, and in sunlit mature tree canopies had average leaf life spans of 3.2, 1.9, and 1.6 years, respectively. Species native to caatinga forest (poorer soils) had average leaf life spans of 4.2, 3.4, and 2.5 years, respectively, in these same micro-site types. Two species common in gaps and in disturbed sites had much longer leaf life span in shaded understory locations than in open, disturbed micro-sites.

The R/FR ratio found in the forest floor is key in determining many important biological processes, as, for example, breaking of dormancy of certain seeds and provoking their germination. When a canopy gap opens due to forest disturbance, the changes in the R/FR ratio may provoke germination of seeds of pioneer species such as *Cecropia* that are in the seed bank in the forest floor (Vázquez-Yanes and Orozco-Segovia 1990).

Understory plants have low photosynthetic capacity and low saturation levels, while canopy trees generally are not saturated, and often have higher photosynthetic capacity. The shaded understory of a tropical forest presents a particular challenge for photosynthetic acquisition of sufficient energy and carbon to support growth and survival (Chazdon et al. 1996). Light in the shaded understory is depleted of PAR and has a low R/FR ratio (i.e., there is lots of FR, because leaves absorb the red and do not absorb the FR). Leaves in the understory are adapted to low flux densities and also to responding very fast to changes in light availability, especially to sunflecks. The sunflecks, or light that passes through holes in the canopy, have similar spectral quality as direct sunlight plus some additional wavelengths from light reflected by leaves. The light intensity and quality of sunflecks thus vary with time of day, angle of the sunfleck, colors of leaves and other plant parts. Sunflecks are less intense than direct light, but between two and five times more intense than full shade. Individual sunflecks usually last less than 2 min.

Carbon assimilation in the understory is strongly dependent on sunflecks, which can provide 30–40% of the daily PPFD (Chazdon et al. 1996). In microsites where sunflecks are more abundant, they could be expected to account for an even larger fraction of the daily carbon gain of understory plants. Shade-adapted plant species have photosynthetic capabilities such as induction that allow them to respond very quickly to sunflecks. Induction means that if a leaf has been exposed to light it will respond faster to the next coming sunfleck (Chazdon and Pearcy 1991).

2.5.3.3 Light Distribution in the Forest

Light distribution varies vertically and also horizontally depending on openness of forest, season of the year, and size of canopy openings or gaps. The upper tree layer of a forest receives 25–100% of relative illuminance depending on the angle of the sun, which in turn depends on latitude and time of day. This is called the euphotic layer (Longman and Jenik 1987). Lower down, below the middle tree layer there is an oligophotic phase with 1–3% and even less illuminance, where seedlings and saplings grow, along with ferns and mosses. Cloud conditions also influence light distribution in the forest. More PAR reaches the understory of a rain forest on an overcast than on a sunny day because the diffuse light of the cloudy day enters the canopy at many different angles, and thus has a greater probability of penetrating the canopy.

Openings in the forest canopy dramatically affect the light environment. Forest canopy openings can be small (selective logging, branch or tree fall) or extensive (landslide, hurricane damage, or land clearing) (Chazdon et al. 1996). Chazdon and Fetcher (1984) compared the light environment in a small (200-m^2) gap, in another gap of twice that size (400 m^2), in a large clearing, and in the understory of the rain forest at La Selva Biological Station in Costa Rica. Total PPFD of the 400-m^2 gap was 20–35% of that in the clearing, and in the 200-m^2 gap, it was 10% that of the clearing. Total PPFD in the center of the 400-m^2 gap was 25 times greater than in the understory, and two to three times greater than in the 200-m^2 gap. The decrease in PPFD in gaps of different sizes was not proportional to size of the gaps.

The influence of a gap may extend far beyond the immediate gap area, as shown by elevated PPFD in understory locations up to 20 m from a gap edge, although this effect has not always been observed (Chazdon et al. 1996). Gap light regimes exhibit substantial yearly variation due to changes in both solar angle and weather conditions. Differences in canopy opening sizes within rain forests promote differences in regeneration survival and growth of canopy tree species (Ashton et al. 1997 a). In Sri Lanka, studies have demonstrated this variability in survival and growth of late-successional canopy species of *Shorea* trees in relation to light, soil moisture, and nutrient status (Ashton and Berlyn 1992; Ashton 1995). Specific management guidelines have thus to be followed to ensure the growth of these species in plantations, enrichment plantings, or other forest management or restoration projects.

2.5.4 Herbivory

The high rate of leaf production and leaf fall in the humid tropics means that there is a relatively large amount of energy that is potentially available to insect herbivores. Leaves are attractive to herbivores and pathogens because they are more nutritious than other parts of the plant. Immature leaves lose area to herbivores at 5–25 times the rate of fully enlarged, mature leaves (Turner 2001). The loss of material to herbivores, mostly insects, can be substantial, and can have a negative influence on tree growth and reproduction (Coley and Barone 1996). For example, Marquis (1987) found that heavy defoliation (50% loss of area or more) of the understory shrub *Piper arieianum* at La Selva, Costa Rica, caused significant reduction in seed output of plants for 2 years subsequent to the leaf loss. Several studies have focused on herbivory in understory palms, perhaps because they have lower stature and are easier to study. Most neotropical palm species seem to be quite resilient to leaf loss and in some cases can even increase reproductive rates following loss of leaf material (Turner 2001).

However, in many tropical forests on nutrient-poor sites, only a small proportion of the available energy stored in leaves is actually utilized by herbivores (Golley 1977). This may be due in part to defense mechanisms in tropical trees, which prevent herbivores from taking advantage of the high quantity of leaves. Many tropical plants produce secondary compounds such as phenolics that make their leaves unpalatable to most herbivores (Levin 1976). However, production of such compounds has an energetic cost (Jordan 2002), and insects may still overcome the defenses (Coley 1980). If the production of defenses uses resources that would otherwise be used to grow more plant tissue, then in certain circumstances there may be selection in favor of maximizing growth, even at the cost of being susceptible to herbivores or pathogens. However, most plants do have some mechanisms of defense against herbivory (Turner 2001).

2.5.5 Decomposition

The rate of organic matter decomposition is called the "*k*" value. It is calculated as follows: if X_0 is the original amount of litter on the forest floor, X is the amount of litter remaining at a later time, e is the base of natural logarithms, and t is the time elapsed between X_0 and X, then

$$X_0/X = e^{-kt} \qquad (2.1)$$

In an ecosystem where rates of litter fall and decomposition rates are approximately equal, and the stock of litter is thus in a steady state, *k* is related to standing stock of litter (X) and rate of litter fall (L) by the equation:

$$k = L/X \qquad (2.2)$$

The time required for 95% of the standing crop to decompose is estimated by $3/k$, and for 99%, $5/k$ (Olson 1963). Comparison of *k* values in various biomes indicates that decomposition is significantly higher in tropical forests (Table 2.2).

Anderson et al. (1983) challenged the conclusion that the decomposition rate of litter is higher in the tropics. They presented a series of *k* values for leaf litter from a variety of habitats and concluded that variation within single regions is too great to permit statements about differences between tropical and temperate latitudes. There certainly is a wide variation in decomposition rates in the tropics. Values of *k* in lowland tropical forests range from

Table 2.2. Decomposition constants *k* and $3/k$ in six ecosystem types. See text for definitions. (Adapted from Swift et al. 1979)

	Tundra	Boreal forest	Temperate deciduous forest	Temperate grassland	Savannah	Tropical moist forest
k year^{-1}	0.03	0.21	0.77	1.5	3.2	6.0
$3/k$ year^{-1}	100	14	4	2	1	0.5

0.76 in a heath forest to greater than 3.0 on soils of intermediate quality (Jordan 1985). However, decomposition follows the same pattern as primary production, and while primary production varies widely in the tropics it is still higher than at high latitudes when the forests that are compared receive similar amounts of rain and are on soils of similar quality. Thus if litter fall is higher in lowland moist tropical forests on rich soils than litter fall in lowland moist temperate forests on rich soils, then decomposition also must necessarily be higher, since, on average, decomposition must equal production. If decomposition in tropical forests were lower than litter fall, there would have to be an ever-increasing stock of organic matter on the forest floor. This is clearly not the case.

It is important to recognize that stocks of organic matter on the forest floor do not indicate the rate of decomposition. Carbon accumulation in soils of the lowland wet tropics ranges from 12–18 kg/m^3 of soil and may be as high as 24 or more in montane forests (Zinke et al. 1984) (organic litter is approximately 50% carbon). These high stocks do not mean that litter is accumulating or that decomposition is slow. Decomposition rate is a function of the amount of *undecomposed* material present (Olson 1963). The greater the amount of organic matter present, the greater the rate of decomposition. A larger biomass of undecomposed material will support a greater mass of decomposers. Thus, as the amount of organic matter increases, the number of decomposers increases, until the rate of decomposition equals the rate of input.

The high rate of organic matter decomposition in tropical forests has important implications for management and conservation (Box 2.4).

Box 2.4

Implications of organic matter decomposition for management and conservation

Most tropical soils are highly weathered (Sanchez 1976) and have a low ability to retain nutrients. Nutrients in rain forest ecosystems are retained primarily in the soil organic matter, where under undisturbed conditions the nutrients are gradually released in a soluble form, or are transferred directly from decomposing litter to roots through mycorrhizae (Herrera et al. 1978). Once a forest or even a small patch of forest is cleared, organic matter input into the cleared area is reduced or eliminated. The litter and humus on the forest floor and the organic matter within the soil quickly disappear. With the disappearance of the soil organic matter, the ability of soil to recycle nutrients is quickly lost, soil fertility rapidly declines, and the ecosystem loses its productive capacity (Jordan 1985). Thus management techniques that preserve litterfall inputs into the soil are especially important in tropical rain forests (Van Wambeke 1992).

For example, in some types of agroforestry systems, leaf litter is used to increase crop yield, especially when leguminous tree leaves are used as mulch. Mulches can protect soils against erosion, decrease weed growth, release nutrients to the soil via decomposition, and moderate soil moisture loss and temperature fluctuations (Montagnini et al. 1993). Farmers frequently use leaf litter as mulch when inorganic fertilizers are too expensive and livestock manure is not available (Byard et al. 1996). Clear-cuts, especially if followed by burning of the accumulated woody debris, are particularly harmful to the ability of the ecosystem to sustain productivity. Therefore, if a tree plantation or agricultural system follows a clear-cut, management has to be geared towards nutrient conservation and replenishment in order to sustain productivity in the long term. More details on nutrient management and conservation in plantations and agroforestry are given in Chapter 6.

2.6 Nutrient Cycling

2.6.1 Cycling Rates in the Tropics

Because energy moves through tropical ecosystems more rapidly than in ecosystems of higher latitude, it might be expected that nutrients also cycle faster in tropical ecosystems, since nutrients follow energy in such functions as leaf production and decomposition. Rates of nutrient cycling can be compared among tropical and temperate forests by taking a systems approach. We can assume that the forest consists of a series of compartments (wood, leaves, litter, soil) each with a stock of a particular nutrient, and then examine the movement of that nutrient in and out of each compartment. The sum of turnover times for all the major compartments in a forest equals the total cycle time for that nutrient. A comparison of turnover times of calcium in various ecosystems indicates that the total cycle time for calcium in tropical forests is shorter than for forests at higher latitudes (Table 2.3). In tropical rain forests, constant warm temperatures and plentiful rainfall throughout the year mean that the processes that govern nutrient cycling never cease. It is the distribution of temperatures throughout the year that differentiates energy flow in the tropics from these processes at higher latitudes (Walter 1971).

Nutrient cycling is higher in forests on fertile soils in the wet tropics than on forests on infertile soils. Studies of nutrient transfer by leaf litter fall in tropical forest ecosystems (Table 2.4) show a large difference in the rate of nutrient transfer through this flux in forests on soils of varying fertility. For

Table 2.3. Compartmental turnover and cycling times for calcium. (Reprinted from Jordan 1995)

Location of ecosystem	Type of ecosystem	Soil	Wood	Canopy	Litter	Total cycle time (years)
Puerto Rico	Rain forest	3.0	6.4	0.9	0.2	10.5
Ghana	Moist forest	8.2	6.8	1.5	0.2	16.7
England	Pine plantation	11.2	6.1	0.8	3.4	21.5
Belgium	Oak-ash	184.9	21.5	0.4	0.6	207.4
Northwest USA	Douglas fir	57.4	20.2	5.5	10.2	93.3
Northeast USA	Northern hardwoods	14.0	10.8	0.8	34.8	60.4

Table 2.4. Production of fine litter and nutrient content of litter in tropical rain forests on soils of different fertility. (Adapted from Montagnini and Jordan 2002)

Type of soil	Litter production kg ha^{-1} $year^{-1}$	Nutrients				
		N	P	K	Ca	Mg
Soils of moderate fertility	10,500	162	8.8	41	171	37
Soils of low fertility	8,800	108	3.1	22	53	17
Oxisol – very low fertility	5,800	61	0.8	6	8	4
Spodosol – extremely low fertility	5,600	42	2.6	27	43	9

example, nutrient cycling on nutrient-poor tropical Spodosols ("caatinga" forest) is slower than in more nutrient-rich soils (Grubb 1995). Extended dry seasons also limit primary production and decomposition. Therefore, annual rates of nutrient cycling would be expected to be lower in tropical dry forests where extended dry seasons occur. For example, in a comparison of productivity, litter fall, and turnover times for tropical forests of the world, the largest litter production was found in tropical moist forest, where it was about three times higher than in tropical dry forests (Brown and Lugo 1982), and, therefore, nutrient fluxes in tropical moist forests also must be much higher than in dry forests.

The variations in the nitrogen cycle between latitudes and between forests on soils of low and moderate fertility are illustrated in Box 2.5. Mechanisms determining the availability of phosphorus in tropical forest soils are shown in Box 2.6.

Box 2.5

The nitrogen cycle in tropical forests

Nitrogen makes up about 78% of the atmosphere, but it can only enter terrestrial ecosystems through nitrogen fixation or via precipitation (Brady and Weil 2002). Nitrogen gas can be fixed from the atmosphere by microbes living symbiotically with the roots of plants in the legume family. It also enters the ecosystem dissolved in rainwater. Once in a plant, nitrogen is used in proteins and other compounds. When a leaf or a plant dies, it falls to the ground as litter. Some of the nitrogen is converted into ammonium and nitrate and moves into the soil or roots. Under anaerobic conditions, such as those that occur on poorly drained or waterlogged soils, some nitrate-nitrogen is reduced to nitrogen gases by bacteria called denitrifiers and is then released into the atmosphere.

The high temperatures and humidity of forests of the moist tropics are near optimum for microbes that mediate nitrogen transformations. As organic matter in the soil decomposes, organic nitrogen is decomposed into ammonia, which can be taken up by plants, or transformed into nitrite and then nitrate by a set of highly specific, aerobic soil bacteria. Nitrate can be leached or taken up by roots or by microbes, or the nitrate nitrogen can be denitrified and returned to the atmosphere (Jackson and Raw 1973).

The microbes that mediate nitrogen transformations are active year-round. As a result, the sparse available data suggest that in most tropical ecosystems, more nitrogen is cycled per year than in temperate zones (Table 2.5). An exception is tropical forests where nitrogen stocks in the soil are extremely low and therefore limiting, as on Spodosols. These are soils that have developed under highly acidic conditions, where organic matter from the topsoil dissolves and precipitates down in the soil profile (Brady and Weil 2002). In the forests near San Carlos de Rio Negro in the Venezuelan Amazon, nitrogen fluxes to the forest floor in unfertile Spodosols were only half the amount of those in Oxisols, which are relatively more nutrient-rich than Spodosols (Cuevas 2001).

Table 2.5. Nitrogen cycling in various forest ecosystems. Pools are given in kilograms of nitrogen per hectare, and fluxes are given in kilograms of nitrogen per hectare per annum. (Adapted from Jordan et al. 1982)

	Amazon rain forest on Oxisol	Amazon rain forest on Spodosol	Seasonal forest, Banco, Ivory Coast	Hardwood forest, Coweeta, North Carolina	Hardwood forest, Hubbard Brook, New Hampshire	Douglas fir, Andrews Forest, Oregon
Pools						
Leaves	143					144
Stems, branches, bark	941	336	1,150	995	351	394
Roots	586	834	–	–	181	197
Litter and superficial humus	406	132	–	140	1,100	798
Soil	3,507	785	6,500	6,917	3,626	3,397
Total	5,583	2,087	7,650	8,052	5,258	4,930
Inputs						
NH_4-N in precipitation	11.3			2.7		
NO_3-N in precipitation	0.2	21.0	21.2	3.6	6.5	2.0
N fixation	16.2	>35	–	12.0	14.2	2.8
Outputs						
NH_4-N leached	8.4	9	21.2	0.06		
NO_3-N leached	5.7			0.1	4.0	1.5
Denitrification	2.9	–	–	10–18	–	–
Balance (input–output)	+8.9			~0	+16.7	+4.3
Internal fluxes						
N in leaf-fall	61.3	24	170	33	54.2	10.8
NH_4-N in throughfall	25.0					
NO_3-N in throughfall	0.3	9	80	4	9.3	3.4
Total internal fluxes	86.6	33	250	37	63.5	14.2

Box 2.6

Availability of phosphorus in tropical forests

Phosphorus is frequently the nutrient element most limiting plant growth in Oxisols and Ultisols (Cuevas 2001), the most common soil types in the tropics (Sanchez 1976; Van Wambeke 1992). These soils have high concentrations of iron and aluminum. Insoluble iron and aluminum phosphates are formed in the mineral soil. Phosphate in this form is unavailable to plants (Brady and Weil 2002). In the undisturbed forest, however, phosphorus availability is not generally a problem. Leaf litter provides an energy source for the microbes that carry out decomposition. As a result of their metabolism, the microbes produce organic acids that react with the iron and aluminum. The organic acids use up the binding power that otherwise would be used to bind phosphorus. The organic acids may also solubilize the phosphorus in iron and aluminum phosphates. Because of this microbial activity, more phosphorus is made available for plant growth, plants grow more, and, as a result, they produce more litter (Jordan 1989). This cycle can be maintained as long as there is a layer of organic matter on the soil. When this layer is destroyed by forest cutting and cultivation, the availability of phosphorus can fall to levels that are below those that are needed for plant growth. Phosphorus is also a limiting factor in Andosols formed from volcanic activity. In humid regions, volcanic ash weathers quickly into allophane, an amorphous aluminum-silicate mixture that rapidly forms complexes with phosphorus, and immobilizes it (Sanchez 1976).

2.6.2 Leaching and Weathering

The high rate of decomposition plus the continuous warm temperatures and high rainfall in the wet tropics result in optimum year-round conditions for leaching of soil nutrients. Positively charged nutrient ions such as calcium and potassium are held on the surface of clay particles, but when the soil solution is acidic, the nutrients are replaced by hydrogen ions. Hydrogen is formed in the soil when carbon dioxide respired by soil organisms reacts with soil water to form carbonic acid, which dissociates into bicarbonate and a positively charged hydrogen ion (Johnson et al. 1977):

$$CO_2 + H_2O \rightarrow H_2CO_3 \tag{2.3}$$

$$H_2CO_3 \rightarrow H^+ + HCO_3^- \tag{2.4}$$

Negatively charged anions such as bicarbonate replace negatively charged nutrient ions such as nitrate and sulfate. The nutrients that have been exchanged into the soil solution can then be adsorbed by soil colloids, used by plants, or leached into streams or groundwater.

The reactions in Eqs. (2.3) and (2.4) also result in soil weathering. Weathering occurs when hydrogen in the soil solution reacts with minerals in the soil or bedrock, resulting in removal of nutrient elements. For example, feldspar is an aluminosilicate (aluminum and silica compound) containing nutrients such as sodium, potassium, and calcium. When it is hydrolyzed, the nutrients are removed from the aluminosilicate.

2.6.3 Nutrient-Conserving Mechanisms

2.6.3.1 "Direct" Nutrient Cycling

Soils in many regions of the lowland wet tropics are very low in nutrient elements because the processes of leaching and weathering are intense and there has been little geologic activity that would result in fresh parent material for the soils. Many soils in humid tropical regions are classified as Ferralsols in the FAO soil classification system (Zech 1993), a term that reflects the iron remaining in the upper soil horizon after the nutrient elements have been removed. Under other systems of classification, such as the US Soil Taxonomy System, Ferralsols are called Oxisols and Ultisols (Brady and Weil 2002; see also Table 2.7). Because of the very low capacity of these soils to retain nutrients (Sanchez 1976) and the high rate of release of nutrients from decomposing litter and soil organic matter, a high rate of nutrient loss from the system might be expected. However, in fact, losses due to leaching in tropical forests are not great and in some cases they can even show a positive balance (Table 2.6). The high Ca losses in Panama shown in Table 2.6 may be due to the exceptionally Ca-rich soil (Golley et al. 1975).

How is it that lowland moist tropical forests are so conservative of nutrients despite the high potential for nutrient loss? One reason appears to be the high concentration of roots and mycorrhizal fungi intertwined with the decomposing litter and humus on or near the soil surface. An example of the role of the root mat in conserving soil phosphorus in a nutrient-poor rain forest is given in Box 2.7.

Table 2.6. Runoff, atmospheric input, and balance of calcium in various ecosystems, arranged in increasing order of calcium runoff. Values are in kilograms per hectare per annum. (Sources of data given in Jordan 1985)

Formation or association and location	Runoff	Input	Balance
Tropical rain forest, Malaysia	2.1	14.0	+11.9
Rain forest on Spodosol, Amazon	2.8	16.0	+13.2
Evergreen forest, Ivory Coast	3.8	1.9	-1.9
Rain forest on Oxisol, Amazon	3.9	11.6	+7.7
Pine forest, North Carolina	4.1	6.5	+2.4
Douglas fir forest, Washington	4.5	2.8	-1.7
Coniferous forest, Minnesota	4.5	3.1	-1.4
Coniferous forest, New Mexico	4.9	7.6	+2.7
Hardwood forest, North Carolina	6.9	6.2	-0.7
Oak-pine forest, Long Island	9.7	3.3	-6.4
Beech forest on sandstone, Germany	12.7	12.8	+0.1
Spruce forest on sandstone, Germany	13.5	12.8	-0.7
Northern hardwoods, New Hampshire	13.9	2.2	-11.7
Tropical rain forest, New Guinea	24.8	0	-24.8
Aspen forest, Michigan	~25	8.3	-16.7
Mixed mesophytic forest, Tennessee	27.4	10.5	-16.9
Montane tropical forest, Puerto Rico	43.1	21.8	-21.3
Tropical moist forest, Panama	163.2	29.3	-133.9

Box 2.7

The role of the root mat in conserving soil phosphorus at San Carlos de Rio Negro in the Venezuelan Amazon

In an Amazon forest on a very nutrient-poor soil (an Oxisol) near San Carlos de Rio Negro, Venezuela, Herrera et al. (1978) showed that nutrients moved directly from decomposing litter through mycorrhizae into the leaf before they could be leached into and through the mineral soil. Herrera et al. (1978) took P^{32}-labeled leaves and placed them on the forest floor. Fine roots of nearby trees grew onto and over the leaves within a matter of weeks. They then harvested the leaves and roots. An electron micro-autoradiograph showed P^{32} in the leaf, the mycorrhizae, and the root hairs, confirming that direct cycling of nutrients from leaves to roots could increase the efficiency of P cycling at this site (Went and Stark 1968). In a follow-up experiment at the same site, Stark and Jordan (1978) showed that when they applied radioactive calcium and phosphorus to the top of a root mat, 99.9% of the activity applied was retained by the roots through a combination of adsorption and uptake. Only 0.1% of the activity leached through the mat of roots, confirming the finding that roots and mycorrhizae prevented most P and Ca leaching in these nutrient-poor Amazon soils.

Table 2.7. Relationships between tropical regions, naturally occurring vegetation, and soils (three systems of classification). (Reprinted from Jordan 1985)

Region	Vegetation	Soil characterization	Great soil group	FAO classification	US Department of Agriculture classification
Lowland wet tropics	Tropical rain forest, high species diversity	Intensely weathered, silica leached, iron and aluminum oxides dominant	Latosol	Ferralsol	Oxisol
Lowland moist tropics and subtropics	Rain forest and seasonal forest	Highly weathered, but clay minerals common	Red-yellow Podzol	Acrisol	Ultisol
Areas where rock high in bases is present	Depends on rainfall regime	High in bases	Terra Roxa	Luvisol	Alfisol
Lowland wet tropics	Tropical heath forest: "kerangas", Amazon "caatinga"	Intensely weathered, iron and aluminum leached, silica dominant	Podzol	Podzol	Spodosol
Moist and wet mountain regions, volcanically derived	Montane forest, high diversity	Derived from volcanic materials	Andosol	Andosol	Andept subgroup
Mountain valley	Forest or grassland	Rich in organic matter	Chestnut	Kastanozems	Mollisol
Montane region, non-volcanic	Variable, depending on temperature and moisture	Bedrock shallow to deep, depending on climate and erosion	Lithosol	Lithosol	Lithic subgroup
River banks with annual flooding	Flooding interval determines vegetation	Alluvial deposit on river banks	Alluvial	Fluvisol	Fluvent subgroup
Tropical desert and savanna	Desert and savanna	Little chemical weathering, high in bases	Desert, Sierozem	Yermosol	Aridisol
Monsoonal	Monsoon adapted	Swelling-type clays	Grummusol	Vertisol	Vertisol

Most diagrams of nutrient cycles do not reflect the direct transfer from organic matter on the surface through mycorrhizae and decomposers into roots. Nutrients are assumed to move from decaying organic matter into the mineral soil, where they are held until taken up by roots of plants or leached into groundwater or streams. Incorporating the direct pathway in diagrams of tropical forest ecosystems is important, as in Fig. 2.9. The arrows through the soil biota emphasize a critical link in the cycle, a link that can easily be broken.

2.6.3.2 Concentration of Roots Near the Soil Surface

As shown above, the concentration of roots in the topsoil can be an effective mechanism for nutrient cycling and conservation, especially in nutrient-poor ecosystems. In general, root tips, essential for nutrient uptake, tend to be congregated in the upper soil horizons of tropical forests (Odum 1970). However, not all tropical forests are dependent upon a concentration of roots on or near the surface to ensure adequate recycling of nutrients. Only those forests on Oxisols, Spodosols, and occasionally Ultisols regularly show concentrations on or near the surface. The relationship between soils and natural vegetation in various tropical regions is shown in Table 2.7. Along a transect from wet tropical forests to seasonal tropical forests, the intensity of weathering and leaching is less because of less moisture in the soil, and so the quantity of roots on or near the surface tends to be less. For example, the Dipterocarp forest in Malaysia and the moist forest in Panama, both of which receive 2 m of precipitation per year, have fewer roots on or near the surface compared with the San Carlos de Rio Negro forests that receive 3.5 m of precipitation per year (see Table 3.1, Chap. 3).

Kingsbury and Kellman (1997) found that presence of root mats on top of the soil in southeastern Venezuela was indicative of high concentrations of aluminum in the soil rather than of low levels of soil nutrients. The growth of roots on top of the soil, forming a dense root mat, was apparently a negative reaction of the plants to the high concentrations of aluminum, which was presumably toxic to the plants. However, as discussed above, high concentrations of aluminum cause phosphorus to be unavailable to plants: therefore high concentrations of aluminum and low availability of phosphorus are strongly related to each other.

Nepstad et al. (1995) showed that some secondary successional trees in the lowland tropics have tap roots as deep as 8 m. These roots may have an important function in water absorption. They could also be an adaptation to nutrients that have leached deeply into the soil following clearing. Their finding does not negate the idea that surficial roots in the wet tropics are important for nutrient transfer.

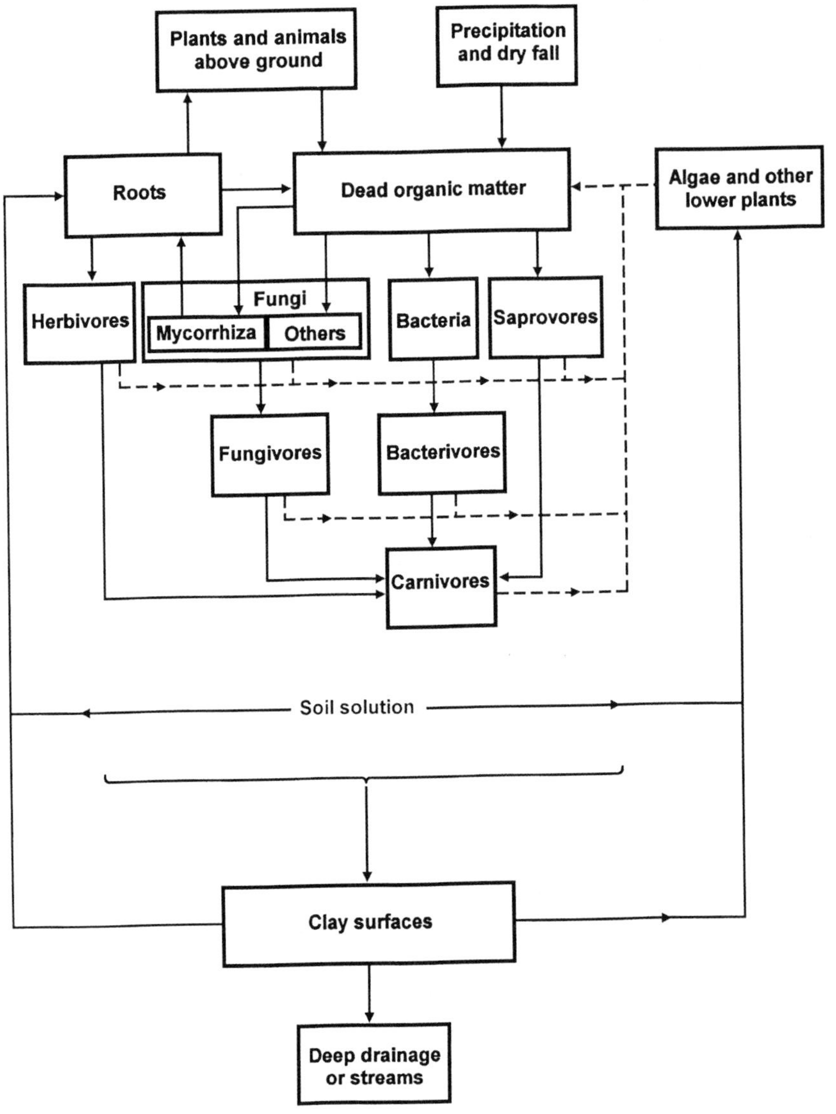

Fig. 2.9. Some of the possible flows of nutrients through the below-ground portion of an undisturbed forest. *Dashed lines* indicate return of organisms to dead organic matter pool. Soluble nutrients may be excreted by many of the organisms, or may be released during trophic transfers. Nutrients in soluble form may not follow trophic pathways. Soluble nutrients may be taken up by algae, or other plants, or they may be leached into the mineral soil where they can be absorbed onto clay, taken up by roots, or leached through the subsoil or to drainage streams. Gaseous fluxes such as nitrogen fixation and denitrification are not indicated. Deforestation greatly simplifies the cycles, and increases the proportion of nutrients in soluble form. (Adapted from Jordan 1985, with permission of John Wiley and Sons Ltd., publisher)

Thick root mats also are common in high-elevation cloud forests. Here the concentration of roots on the surface occurs because of low nitrogen availability (Tanner et al. 1998). Nutrients are tied up in a thick organic mat that has formed because the anaerobic conditions of the saturated environment have slowed decomposition (Bruijnzeel and Veneklaas 1998).

Surface concentrations of litter and humus permeated with fine roots are not confined to the tropics. In old-growth stands of beech (*Fagus grandifolia*) on Ultisols in southeastern USA, there is often a thick mat where roots are in direct contact with decaying leaves (Jordan, pers. observ.). Beech is a climax species and casts heavy shade. Decomposition apparently is slow beneath the canopy and nutrient availability in the decomposing litter may be higher than in the mineral soil below.

2.6.3.3 Nutrient Storage in Wood Biomass

Early researchers who found very low concentrations of nutrients in highly weathered soils of the lowland tropics suggested that nutrients were stored primarily in the above-ground biomass of the trees and that that was how nutrients were conserved (Richards 1952). However, a survey that compared nutrient stocks in above-ground biomass in a variety of tropical forests indicated that only certain nutrients are stored primarily in the living biomass (Fig. 2.10). For example, this is true for calcium and potassium in lowland ecosystems that receive more than 2,000 mm/year precipitation. Both calcium and potassium are highly soluble and they are rapidly leached from soils by heavy rains. For these same ecosystems, organic phosphorus is also primarily stored in aboveground biomass. Upon decomposition, the organically bound phosphorus becomes soluble and available to roots through direct recycling. That which is not recycled leaches down to the mineral soil where it becomes bound by iron and aluminum. Although the total amount of inorganic phosphorus in the mineral soil can be high, little is generally available to trees because of the fixation by iron and aluminum. The pattern for nitrogen is different. Relatively little is stored in the trunk of trees. Most is in the organic matter of the soil (Jordan 1985).

Because most of the cations in tropical forests on nutrient-poor soils are stored in the above-ground biomass or the soil organic matter, clear-cutting and removal of the native forest can result in loss of a large portion of the nutrient stocks, even if the area is immediately replanted with trees (Fölster and Khanna 1997).

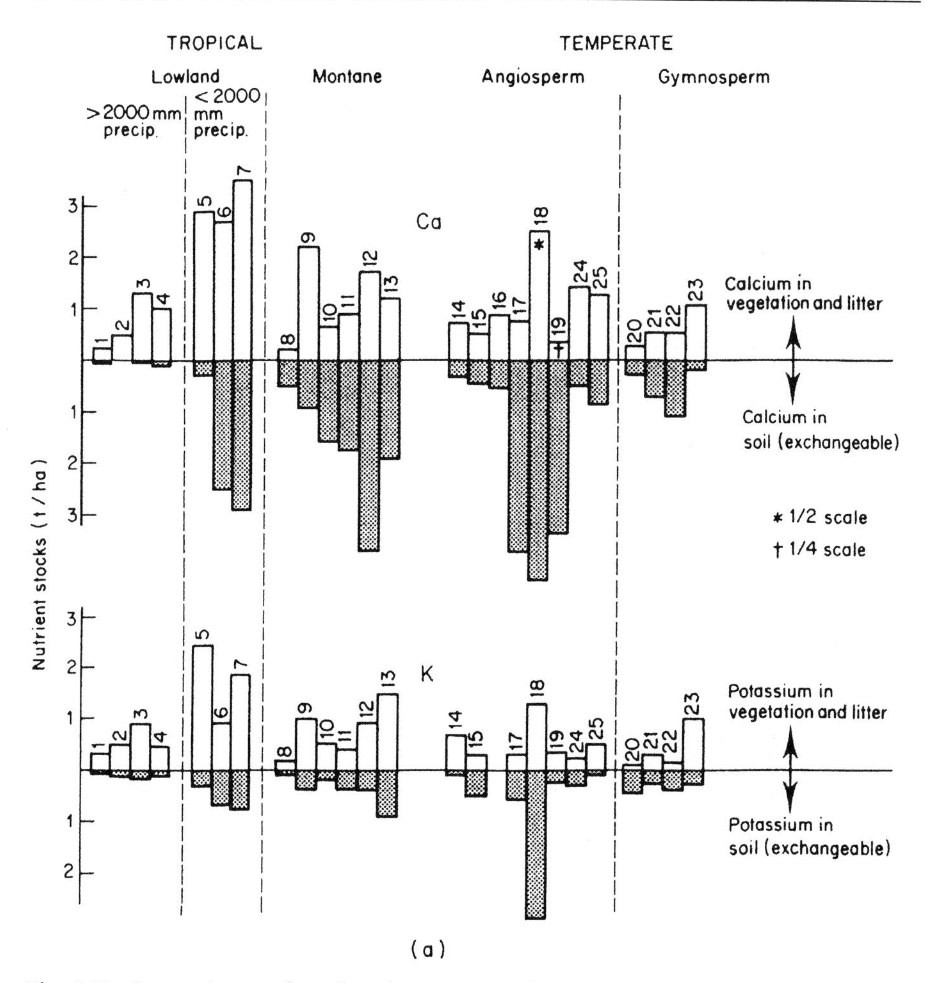

Fig. 2.10. Comparisons of stocks of nutrients in biomass (living vegetation plus litter) and soils of tropical and temperate forests: **a** calcium and potassium; **b** phosphorus and nitrogen. The sites are: *1* rain forest, Venezuela; *2* rain forest, Manaus, Brazil; *3* moist forest, Pará, Brazil; *4* Banco plateau forest, Ivory Coast; *5* mature forest, Thailand; *6* 40-year-old forest, Ghana; *7* seasonal forest, Venezuela; *8* montane forest, Puerto Rico; *9* rain forest, Costa Rica; *10* montane forest, New Guinea; *11* slope forest, Colombia; *12* montane forest, New Guinea; *13* montane forest, Venezuela; *14* beech forest, Sweden; *15* beech forest, Germany; *16* hardwoods, New Hampshire, USA; *17* yellow poplar, Tennessee, USA; *18* oak forest, Oklahoma, USA; *19* oak forest, Virelles, Belgium; *20* 30-year-old pine forest, Ontario, Canada; *21* Douglas fir forest, Washington, USA; *22* spruce-fir, British Columbia, Canada; *23* montane fir, Washington, USA; *24* mixed oak, Tennessee, USA; *25* aspen-maple, Wisconsin, USA. (Sources of data are given in Jordan 1985, pp 40–41. Adapted from Jordan 1985, with permission of John Wiley and Sons Ltd., publisher)

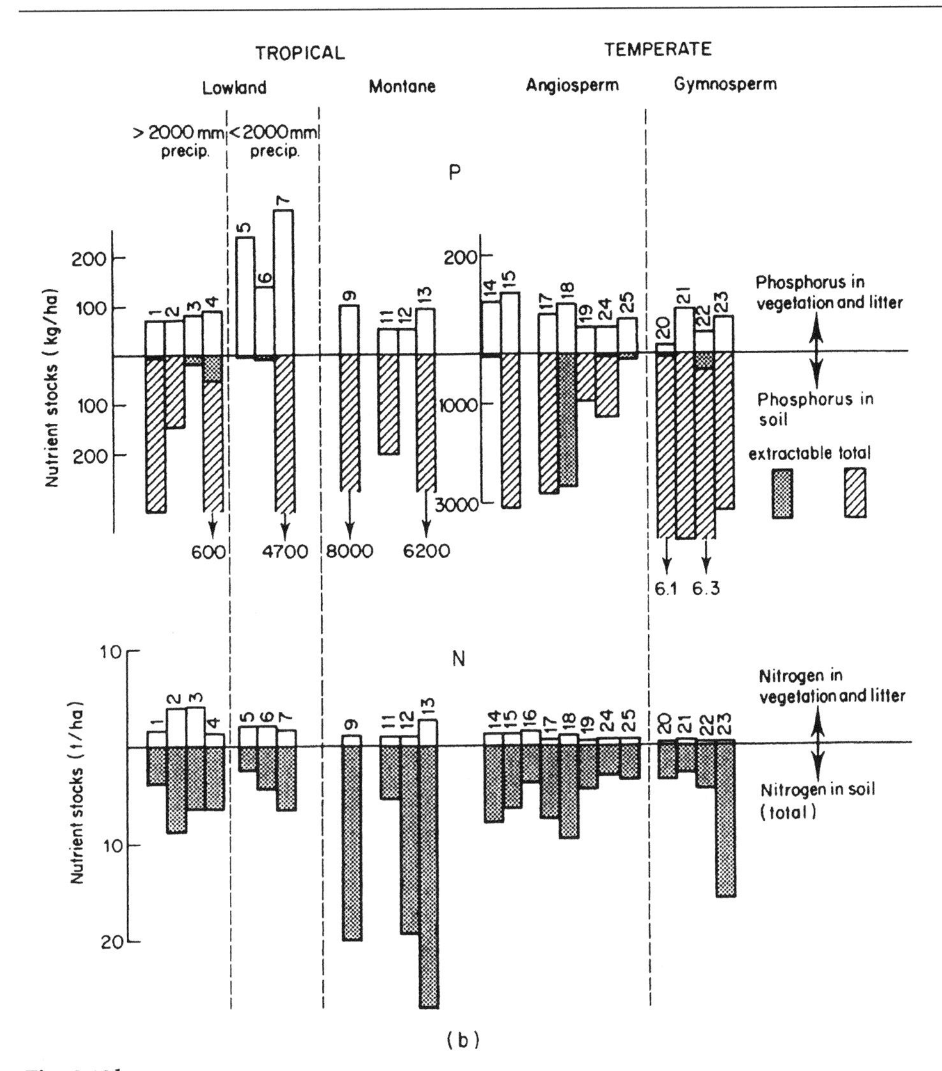

Fig. 2.10b

2.6.3.4 Other Nutrient-Conserving Mechanisms

Direct transfer of nutrients from litter to roots and root concentration near the soil surface are only two of several mechanisms that have evolved in tropical forests to conserve nutrients (Montagnini and Jordan 2002). Others include:

- *Aerial roots.* Mats of living and dead bryophytes, lichens, club mosses, bromeliads, ferns, orchids, and other epiphytes often occur on the branches and stems of rain forest trees. Adventitious tree roots penetrate the mats. Morphological evidence for the role of such roots in nutrient transfer includes abundant root hairs, unsuberized root tips, and the presence of endomycorrhizal hyphae (Nadkarni 1981).
- *Rapid soil drainage.* The high concentration of soil organic matter near the surface of tropical rain forests results in a relatively stable aggregation of soil particles. Water can drain rapidly through the pores that exist between these stable aggregates, and there is limited contact between soil particles and drainage water, thus limiting leaching potential (Nortcliff and Thornes 1978). When forests are cleared, the soil organic matter breaks down, the aggregates lose their stability, and the soil becomes less permeable. As a result, there is more runoff and erosion.
- *Scleromorphic leaves.* Evergreen scleromorphic leaf types are commonly associated with nutrient-poor tropical forests (Medina 1995). The young leaves of tropical evergreen species may be susceptible to insect attack, but as they mature they become thick and tough. Their resistance to herbivory, fungal infection, and attack by other predators and parasites increases (Coley 1982, 1983). Such leaf characteristics may be advantageous to plants in nutrient-poor areas where leaf replacement is energetically expensive (Chapin 1980). In many species, mobile nutrients such as phosphorus, nitrogen, and potassium are translocated from leaf to twig before leaf abscission (Charley and Richards 1983).
- *High nutrient use efficiency in leaves.* Leaves in tropical forests tend to be low in nutrients, and thus their "nutrient use efficiency" is high. In other words, the trees are able to synthesize leaves with limited nutrient availability (Vitousek 1982, 1984). In most lowland forests leaves tend to use phosphorus efficiently, whereas in Spodosols and in montane forests, nitrogen use efficiency tends to be highest.
- *Epiphylls.* Leaves in the humid tropics are often covered with epiphylls such as mosses, lichens, and algae. Some of the lichens and algae fix nitrogen (Forman 1975). Almost all epiphylls appear to be effective in scavenging nutrients from rainwater (Jordan et al. 1979).
- *Secondary compounds in plants.* Secondary plant compounds are those compounds in plants that do not play a role in the metabolism of the plant. These secondary compounds may serve as chemical defenses against pathogens and herbivores. The concentration of secondary compounds in plants of nutrient-poor regions is often high, because replacement of nutrients to synthesize new leaves is energetically expensive (Janzen 1974).

Some nutrient conserving mechanisms are not unique to the tropics. Monk (1966) suggested that the sclerophyllous needles of pine trees at high as well as at low latitudes conserve nutrients. They are long-lived, and resistant to leaching, insect attack, and decay. As mentioned before, surficial root mats of decaying litter interspersed with roots, important in low-nutrient forests of the tropics, also occur in mature beech stands on Ultisols in the Piedmont of southeastern USA (Jordan, pers. observ.).

2.6.3.5 Role of Soil Organic Matter in Nutrient Conservation

Conservation of soil organic matter is important for sustainable management of tropical forests (Van Wambeke 1992). Soil organic matter provides additional exchange sites for cations in the soil, thereby decreasing leaching potential. Organic nitrogen is bound in the soil organic matter and is released more in synchrony with plant needs than it would be as nitrates or ammonia, the common forms of inorganic nitrogen in mineral soil (Brady and Weil 2002). Soil organic matter also makes the soil less susceptible to erosion. Organic compounds synthesized by soil microorganisms help bind clay particles together into aggregates that render the soil more permeable to water (Stockdale et al. 2001). Soil organic matter is also the source of energy for microorganisms whose activity renders the soil more permeable to roots (Jordan 1998).

Because of the importance of soil organic matter for recycling cations, maintaining phosphorus availability, sequestering nitrogen, and preventing soil erosion, management activities should emphasize the maintenance of the litter and humus on the soil surface. Clear-cuts are especially deleterious because the increase in solar radiation reaching the forest floor results in a quick decomposition of the litter. Fire is even more damaging to tropical forest ecosystems when it destroys the top layer of litter and soil organic matter. Fire can instantaneously transform the phosphorus from an organic form in the leaf, where it is slowly released in synchrony with the demand by plants, to an inorganic form where it is bound up and made insoluble by iron and aluminum. Fire converts calcium and potassium to easily leached ash, and volatilizes the nitrogen in tree biomass.

While all fires destroy soil organic matter, the organic matter lost in small-scale shifting cultivation can be quickly replaced by litterfall and tree fall from surrounding forest. However, large-scale burning, as is used to maintain pasture in the Amazon region, can result in an almost permanent loss of organic matter, and consequent long-term loss of productivity (Buschbacher 1986).

2.6.4 Effects of Disturbance on Nutrient Stocks in the Soil

To try to better understand nutrient dynamics following disturbance in the tropics, an 8-year study of nutrient dynamics before and after slash-and-burn agriculture was carried out in the lowland rainforest near San Carlos de Rio Negro, in the Venezuelan Amazon (Jordan 1989; Montagnini et al. 2000). While other types of disturbances in tropical forests have different intensities, sizes, and durations, and the quantities of nutrient loss and recovery are different, the overall pattern of nutrient dynamics may be similar, as evidenced by studies of nutrient dynamics in tropical pastures presented later in this section.

Shifting agriculture has been practiced in the tropics for many centuries. Shifting (also called "swidden", "slash-and-burn") agriculture is the predominant land-use practice on a large portion of the arable soils of the world and provides sustenance for many millions of people (Andriesse and Schelhaas 1987). Traditional shifting agriculture uses long forest fallows between short periods of farming. Long fallows make the traditional technique sustainable but also require extensive amounts of land. When land is scarce, farmers shorten forest fallows and lengthen agricultural periods, resulting in soil nutrient depletion, reduced crop yields, and increased weed invasion. Similar patterns are reported in the tropics worldwide, in spite of differences in ecological and socioeconomic conditions.

There have been many studies to try to explain decreases in productivity of shifting agriculture after relatively short periods of farming. There have been at least two major hypotheses to explain the sharp decreases in growth of crops like corn and rice after 2 or 3 years of shifting cultivation. One is that the decline in crop yield is due to competition from vigorous weeds that have invaded the site (Uhl et al. 1982). The other is that the decline is caused by nutrient losses due to leaching of calcium and potassium and volatilization of nitrogen (Nye and Greenland 1960). A third hypothesis combines the first two: there is substantial nutrient loss, but weedy species are more effective in scavenging low levels of nutrients in the soil.

Average temperature at the San Carlos site was 26 °C and the soil was classified as an Oxisol. Total standing stocks of calcium, potassium, nitrogen, and phosphorus were measured in the undisturbed forest and in the soil. Then an area of about a hectare was cut and burned by local farmers, and planted with manioc (*Manihot esculenta*), plantain (*Musa paradisiaca*), and cashew (*Anacardium occidentale*). The plot was farmed for 3 years, and then abandoned. Total stocks in the crops and weeds (successional vegetation) were measured yearly. Production of manioc roots declined from 1,465 kg/ha in the first year to 700 kg/ha in the third year. Biomass of the other crops was negligible, but biomass of weeds increased from 300 kg/ha in the first year to 990 kg/ha in the third.

Nutrient stocks at the experimental site as a function of time are shown in Fig. 2.11. In the undisturbed forest, most of the calcium and potassium was held in the biomass of the above-ground vegetation. Cutting and burning of the biomass converted most of the wood to ash, and, as a result, much of the calcium and potassium was quickly leached into the mineral soil and held temporarily on the surface of clays in an "exchangeable" form (easily replaced by hydrogen ions). The amounts in the mineral soil gradually decreased, and a year after abandonment, levels were close to those in the undisturbed forest. However, the total amounts in the soil at the time of abandonment were still greater than the total amounts in the crops during the first 2 years of cultivation, and were greater than the amounts in the soil of the undisturbed forest. Therefore it is unlikely that a shortage of calcium and potassium caused the decline in crop production.

The dynamics of nitrogen was different from the cations. Since most of the nitrogen is held in the organic matter of the soil, it was little affected by the cut and burn. The cultivation of the plot may have resulted in oxidation of the soil organic matter with a consequent loss of nitrogen, due to volatilization (Vitousek 1981). However, the quick recovery of nitrogen in the soil stocks during the third year of cultivation suggests a statistical anomaly. In any case, it would seem that lack of nitrogen was not a cause of decline of crop productivity.

In the undisturbed forest, over 83% of the 300 kg/ha stock of phosphorus was in the soil. However, little was available for uptake by roots, because most of it was bound by the iron and aluminum in the mineral soil. When the burned remains of the trees and the decomposing organic matter on the soil surface were almost gone, and crop production had declined, the local farmers gathered leaf litter from the surrounding forest and piled it around the stems of the manioc. As a result, plants resumed vigorous growth. An analysis of soil samples for phosphorus using a fractionation method that separated available from total stocks (Potter et al. 1991) was carried out. Soils were collected from undisturbed sites, from sites where leaf litter had been piled on bare soil, and from bare soil. Results showed that available forms of phosphorus were much higher where soil organic matter was present. A probable explanation is that organic acids leach from decomposing organic matter and react with iron and aluminum in the soil, chelating them, and thereby preventing them from reacting with phosphorus (Ae et al. 1990). As a result, phosphorus remains soluble and available to plant roots.

The conclusion was that the decline in crop productivity was due to neither nutrient loss through leaching and volatilization nor weed competition. The productivity decline was rather due to the fixation of phosphorus by iron and aluminum, following the disappearance of organic matter on and in the soil. Addition of more organic matter to the soil, as done customarily by the local farmers, resulted in leaching of organic acids, which in turn kept phosphorus available.

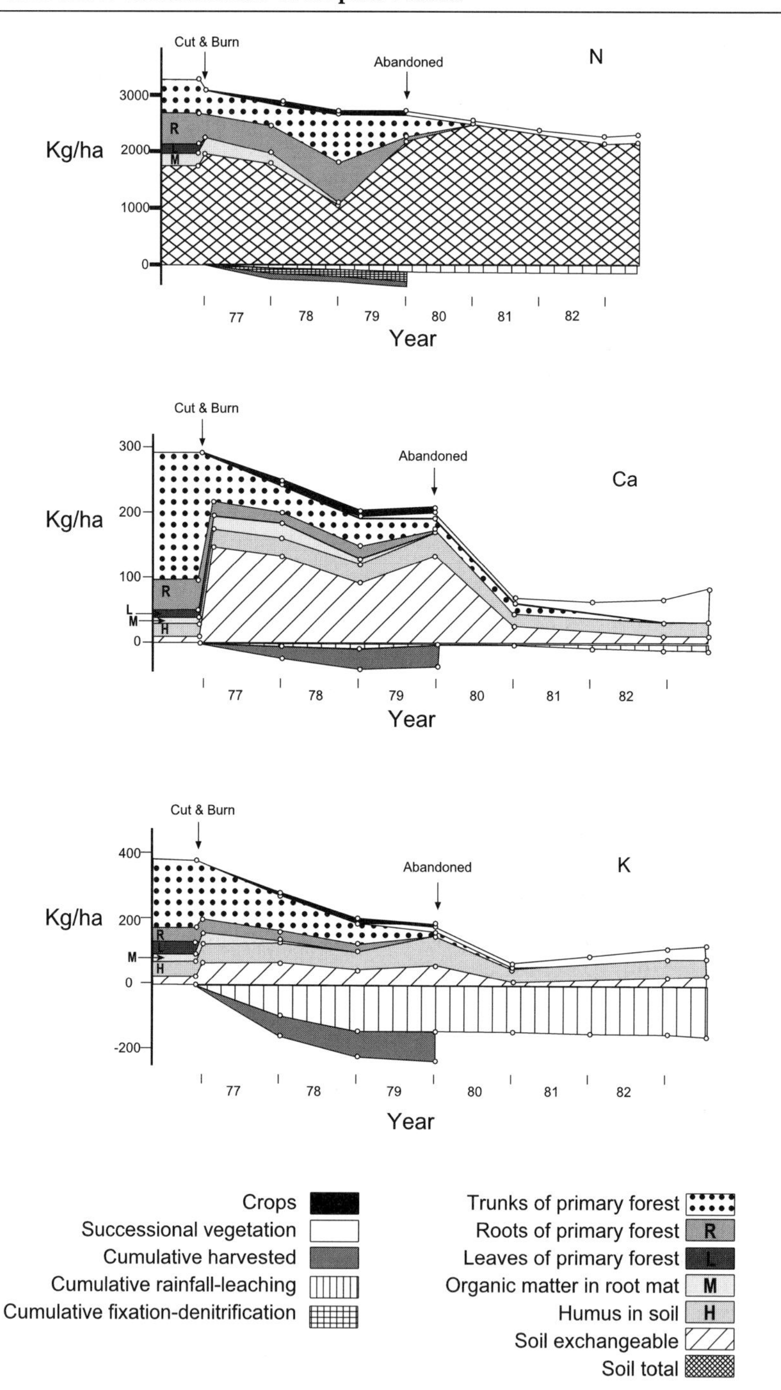

Fig. 2.11. Stocks and cumulative losses of potassium, calcium, and nitrogen as a function of time in the experimental plot at San Carlos de Rio Negro, Venezuela. (Adapted from Jordan 1985, with permission of John Wiley and Sons Ltd., publisher)

Studies in other regions of the Amazon Basin following conversion of forest to pasture suggest a similar pattern (Fig. 2.12). When phosphorus is supplied to the pasture, productivity can remain relatively high (Fig. 2.13), but if it is not, phosphorus availability declines (Fig. 2.12). A common practice in the Amazon when pasture production declines is to burn the shrubs that have invaded the pasture, thus adding a small pulse of nutrients to the soil (Hecht 1984). However, if nitrogen is not replenished, the stocks gradually decline. Although stocks of nitrogen are slow to diminish, they are also slow to be replenished following abandonment of the pasture.

Because of the low nutrient-holding capacity of the soils at San Carlos de Rio Negro, the time scale of the nutrient dynamics may not be typical for all lowland rain forests. Nevertheless, the pattern should not vary. Potassium, a

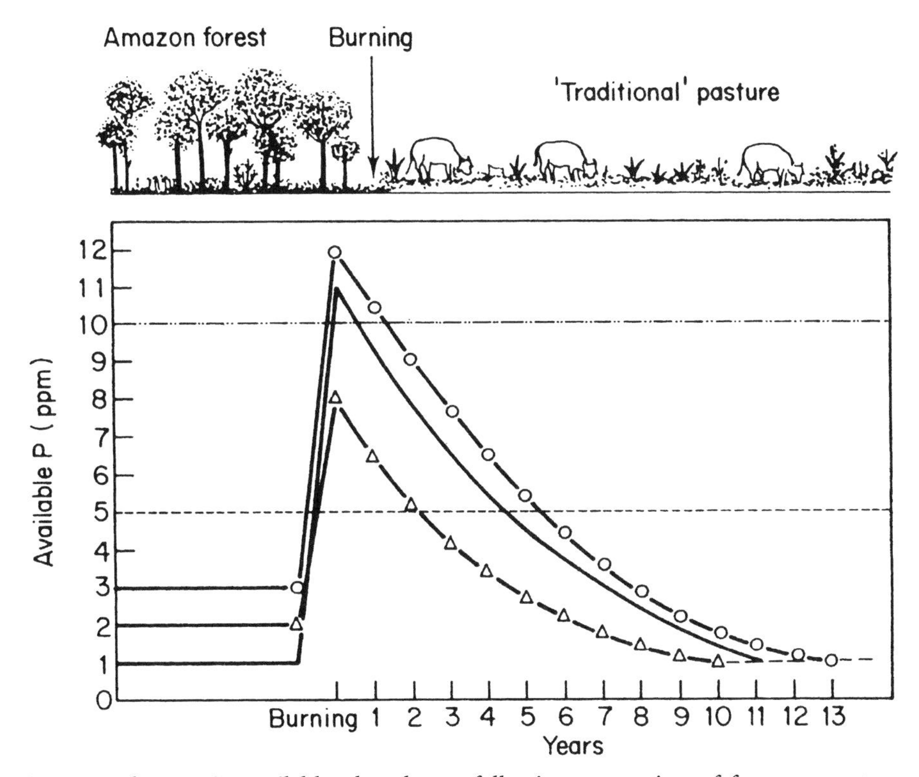

Fig. 2.12. Changes in available phosphorus following conversion of forest to pasture in the Amazon region of Brazil on various soil types: *line without symbols* yellow Latosol (Oxisol), very heavy texture; *line with circles* red-yellow podzolic (Ultisol), medium texture; *line with triangles* dark red Latosol (Oxisol), medium texture. Levels of P for sustainable pasture: *dotted* and *dashed line* standard level; *dashed line* critical level. (Adapted from Jordan 1985, with permission of John Wiley and Sons Ltd., publisher)

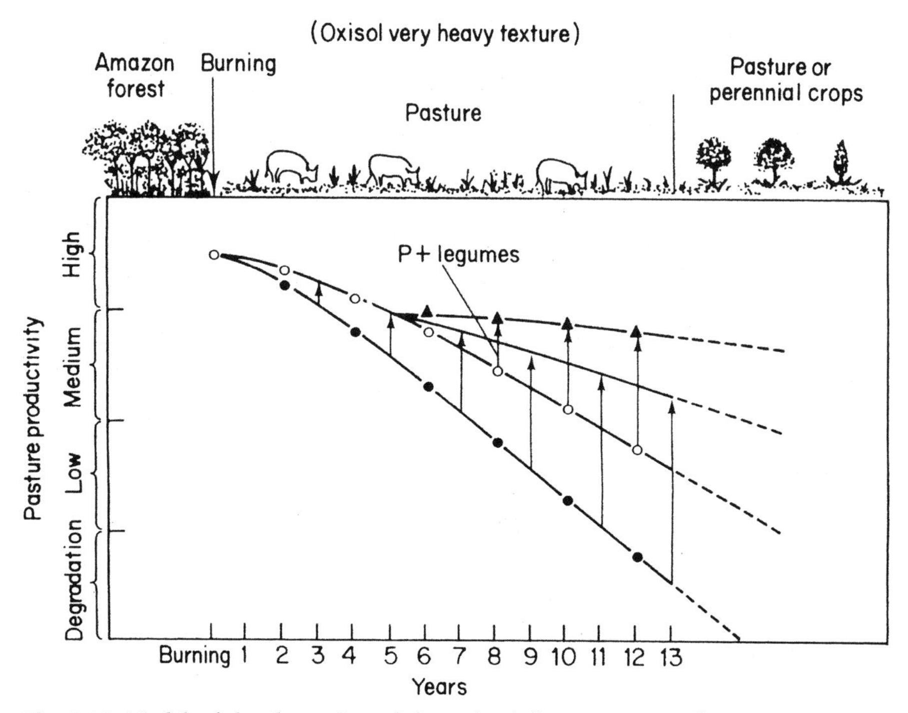

Fig. 2.13. Model of the dynamics of the animal–forage grass–soil system in an Amazon Oxisol, with and without the addition of phosphorus and nitrogen-fixing species. *Line with open circles* Traditional pasture at optimum grazing pressure; *line with solid circles* traditional pasture at grazing pressure above optimum; *line with triangles* improved pasture at optimum grazing pressure; *solid line* improved pasture at grazing pressure above optimum. Improved pasture means grass and legume plus phosphorus; traditional pasture means grass (*Panicum maximum*). (Adapted from Jordan 1985, with permission of John Wiley and Sons Ltd., publisher)

monovalent cation, is easily leached, and will decrease rapidly. Calcium will be also leached but more slowly since it is a divalent cation and is held more firmly in the soil. Nitrogen, in contrast to other nutrients, has its greatest concentration in the soil. Clearing and burning of the aboveground biomass have little immediate effect on stocks of nitrogen in the soil. It is lost slowly, but once it is gone, it is slow to recover. Phosphorus is often a critical element, and its loss due to fixation will vary with soil type (Brady and Weil 2001).

2.6.4.1
Implications for Forestry

Logging of tropical forests may or may not cause nutrient losses as severe as those depicted in Fig. 2.11, depending on the severity of the disturbance to the forest structure and the soil organic matter. Conversion of native forests to plantations may result in a greater loss than that resulting from selective logging, since conversion to plantations involves eliminating the entire forest, along with the nutrient-conserving mechanisms. However, losses may not be as severe as those following agriculture, because once the plantation species are planted, further disturbance is often minor compared to agriculture.

2.7
Conclusion

The most important ecological characteristics of tropical forests to be considered in management are:

- high diversity of tree species
- high frequency of cross-pollination
- common occurrence of mutualisms
- high rate of energy flow through primary producers, consumers, and decomposers
- a relatively "tight" nutrient cycle.

Logging, as well as other types of forest disturbance brought about by management, seriously affects these characteristics and can threaten the ability of a tropical site to regenerate a forest. It is for this reason that tropical forests are often considered fragile (Farnworth and Golley 1974). However, there are means of managing tropical forest that are less disruptive to the forest structure and function. Some techniques for managing tropical forest lands, such as agroforestry, even take advantage of nutrient cycling characteristics of the natural forests of the areas where they are practiced. Management techniques that take into consideration these characteristics of tropical forests are reviewed in Chapters 5 and 6.

Classification of Tropical Forests 3

3.1
Classification Based on Forest Structure

The idea that forest structure reflects productive potential led the 19th century naturalists and explorers to believe that tropical forest regions have great potential for wood and crop production. They were impressed with the huge structure and impenetrable vegetative growth of rain forests (Jordan 1982). In their homelands, such structure and growth indicated a high productive potential. However, in many tropical forests, structure does not reflect function. For example, in the Luquillo Mountains of Puerto Rico, the Colorado forest (*Cyrilla racemiflora*), which dominates between 600 and 800 m above sea level, has a tree density and basal area greater than the Tabonuco forest (*Dacryodes excelsa*) which grows below 600 m elevation (Lugo and Scatena 1995). However, the net primary productivity of the Tabonuco forest averaged 4.86 Mg ha^{-1} year^{-1} (Jordan 1971b), while that of the Colorado forest averaged only 0.59 Mg ha^{-1} year^{-1} (Weaver 1995). In the Amazon region, the structure of forests on nutrient-poor soils also does not reflect the productive potential of the site. In spite of growing on poor soils, the forest has biomass (above ground plus roots) in the range of 320 to 400 Mg ha^{-1} (Jordan 1985), values that are comparable with other humid tropical forests growing on a variety of conditions (324 Mg ha^{-1}, Achard et al. 2002). Nevertheless, above-ground wood production on the nutrient-poor sites was 3.9 to 4.9 Mg ha^{-1} year^{-1}, compared with a value of about 7 Mg ha^{-1} year^{-1}, the average for all mesic tropical forests (Fig. 2.6).

Management for wood production based upon structure is satisfactory when structure reflects function. Structure reflects function when moisture, for example, is a critical factor. Figure 3.1 shows that forests in arid climates differ greatly in wood biomass, leaf biomass, height, and basal area from forests in moist climates. In the case of nutrient status, structure does not always reflect function. Nutrient-conserving mechanisms can compensate for low nutrient availability, with the result that the biomass of a forest on nutrient-poor soils differs only slightly from that of a forest on rich soils (Table 3.1), despite the fact that wood production on the two sites differs greatly.

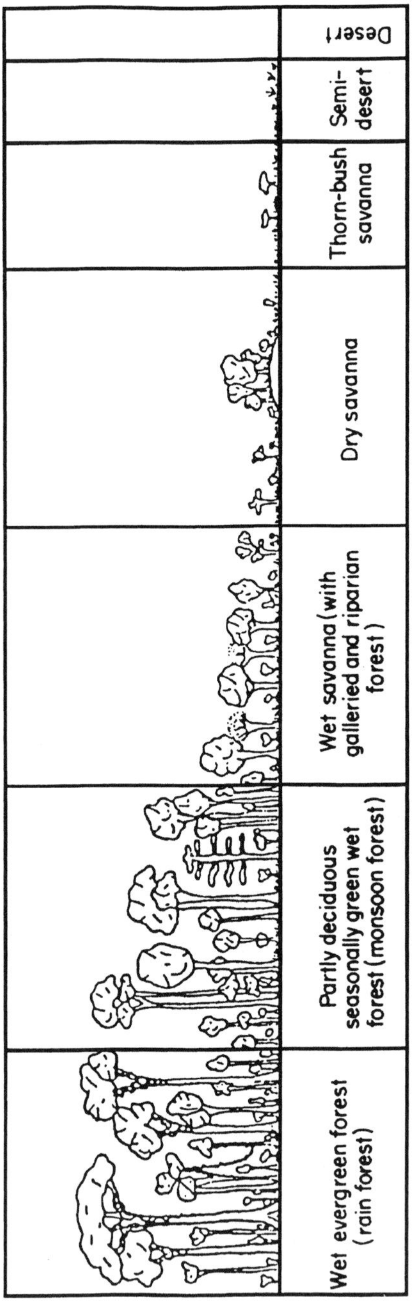

Fig. 3.1. Structure of tropical vegetation along a moisture gradient. (Adapted from Jordan 1985, with permission of John Wiley and Sons Ltd., publisher)

Table 3.1. Ecosystem characteristics along a nutrient gradient. (Adapted from Jordan 1985)

Parameter	Forest type						
	Amazon caatinga, San Carlos, Venezuela	Forest on Oxisol, San Carlos, Venezuela	Lower Montane, El Verde, Puerto Rico	Evergreen, Banco, Ivory Coast	Dipterocarp, Pasoh, Malaysia	Lowland, La Selva, Costa Rica	Forest on dolomite, Panama
1. Root biomass (t/ha)	132	56	72.3	49	20.5	14.4	11.2
2. Aboveground biomass (t/ha)	268	264	228	513	475	382	326
3. Root:shoot ratio	0.49	0.21	0.32	0.10	0.04	0.04	0.03
4. Percentage of roots in superficial mat	26	20	~0	~0	~0	~0	~0
5. Specific leaf area (cm^2/g)	47	65	61	–	88	139	131–187
6. Leaf area index	5.1	6.4	6.6	–	7.3	–	11–22
7. Predicted leaf biomass (t/ha)	10.8	9.8	10.8	–	8.3	–	10.4
8. Leaf litter production ($t\ ha^{-1}\ year^{-1}$)	4.97	5.87	5.47	8.19	6.30	7.83	11.3
9. Predicted turnover time of leaves (years) (row 7/row 8)	2.2	1.7	2.0	–	1.3	–	0.9
10. Aboveground wood productivity ($t\ ha^{-1}\ year^{-1}$)	3.93	4.93	4.86	4.0	6.4	–	–
11. Leaf decomposition constant, *k*	0.76	0.52	2.74	3.3	3.3	3.47	3.2
12. Biomass:phosphorus ratio in leaf litter	2,631	7,237	5,000	1,365	3,282	2,024	1,319
13. Biomass:nitrogen ratio in leaf litter	135	95	–	64	82	52	–

3.2 Classification Based on Forest Function

Forest function may be a more important factor than structure for managing tropical forests for wood production. The most important functions in tropical forests are those related to diversity and mutualisms, nutrient cycling, and energy flow (Chap. 2).

- A diverse forest highly dependent upon mutualisms for pollination must be managed differently from a forest that is pollinated primarily by wind. If mutualisms are not considered, the remaining trees may not be able to reproduce.
- A forest on nutrient-poor soil, where surficial root mats are important in recycling nutrients, must be managed much less intensively than a forest on nutrient-rich soil. If disturbance to the root mat is not minimized during logging operations, nutrient losses will be critical.
- A forest at high elevations, where energy flow is low (low rates of production and decomposition), must be managed differently from a lowland forest where production and decomposition are high. Logging cannot be as frequent and care must be taken to preserve the undecomposed litter on the forest floor, to minimize loss of nutrients.

Because of the lack of understanding of how tropical forests function, functional characteristics are rarely used as a basis for classification of tropical forests. More commonly, they are classified by the climate in which they occur, by species, by successional stage, or by the soil type on which they occur.

3.2.1 Climatic Classifications

Climatic classifications are generally useful where soil conditions are relatively uniform. A number of climatic classifications have been developed. Holdridge's Life-Zone system of classification (Holdridge 1967) is one of the most generalized and used to classify tropical forests. The system classifies vegetation on a global scale, according to latitudinal regions and altitudinal belts based on biotemperature (the average annual temperature using values only between 0 and 30 °C, the range within which photosynthesis has a net positive value). Each of the temperature regions contains a series of Humidity Provinces, ranging from semi-parched (where the ratio of potential evapotranspiration to rainfall is greater than 32), through humid (where potential evapotranspiration and rainfall are equal) to saturated (a theoretical situation

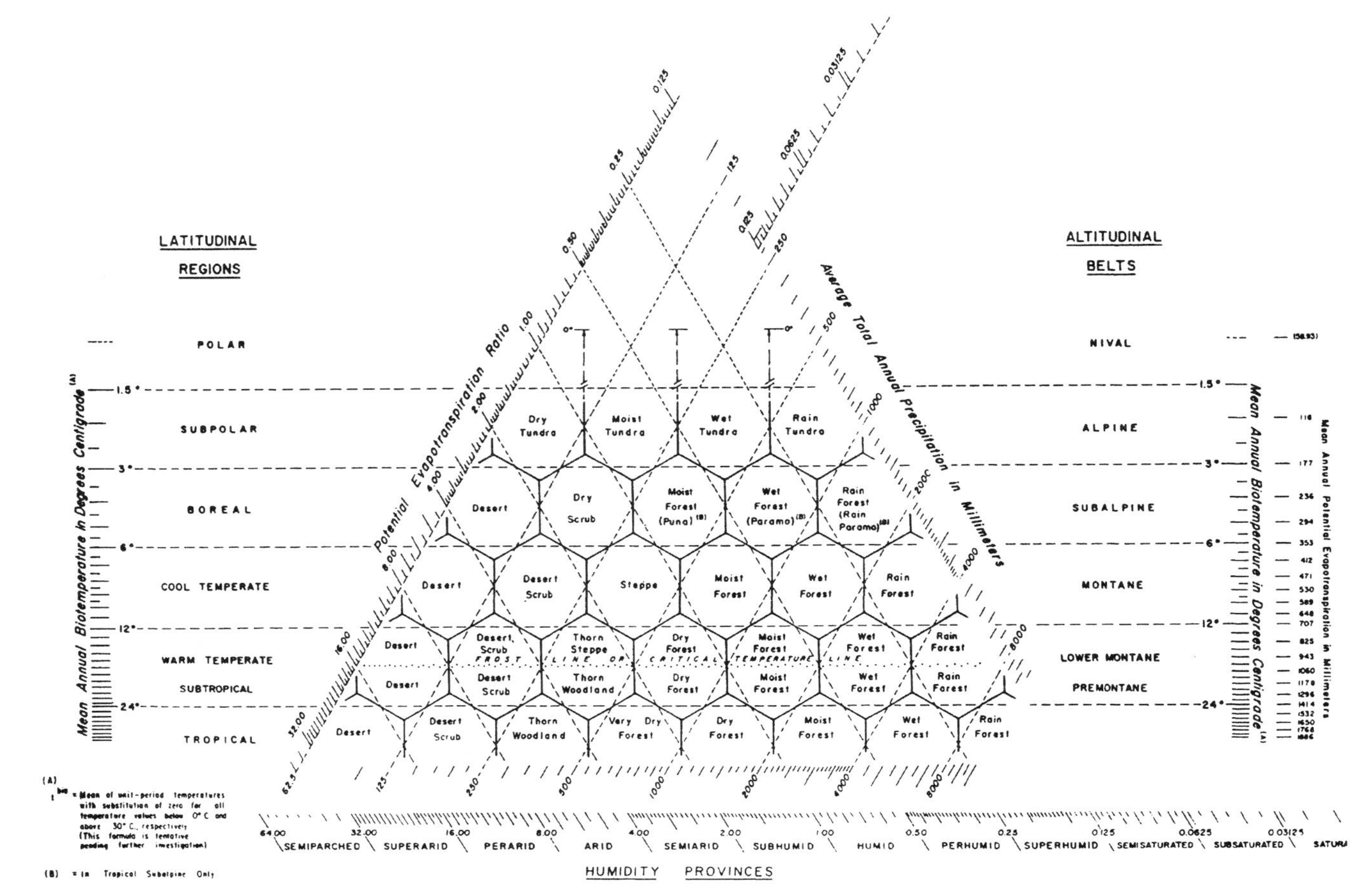

Fig. 3.2. Vegetation types characteristic of various life zones as defined by precipitation and biotemperature. (Adapted from Holdridge 1967)

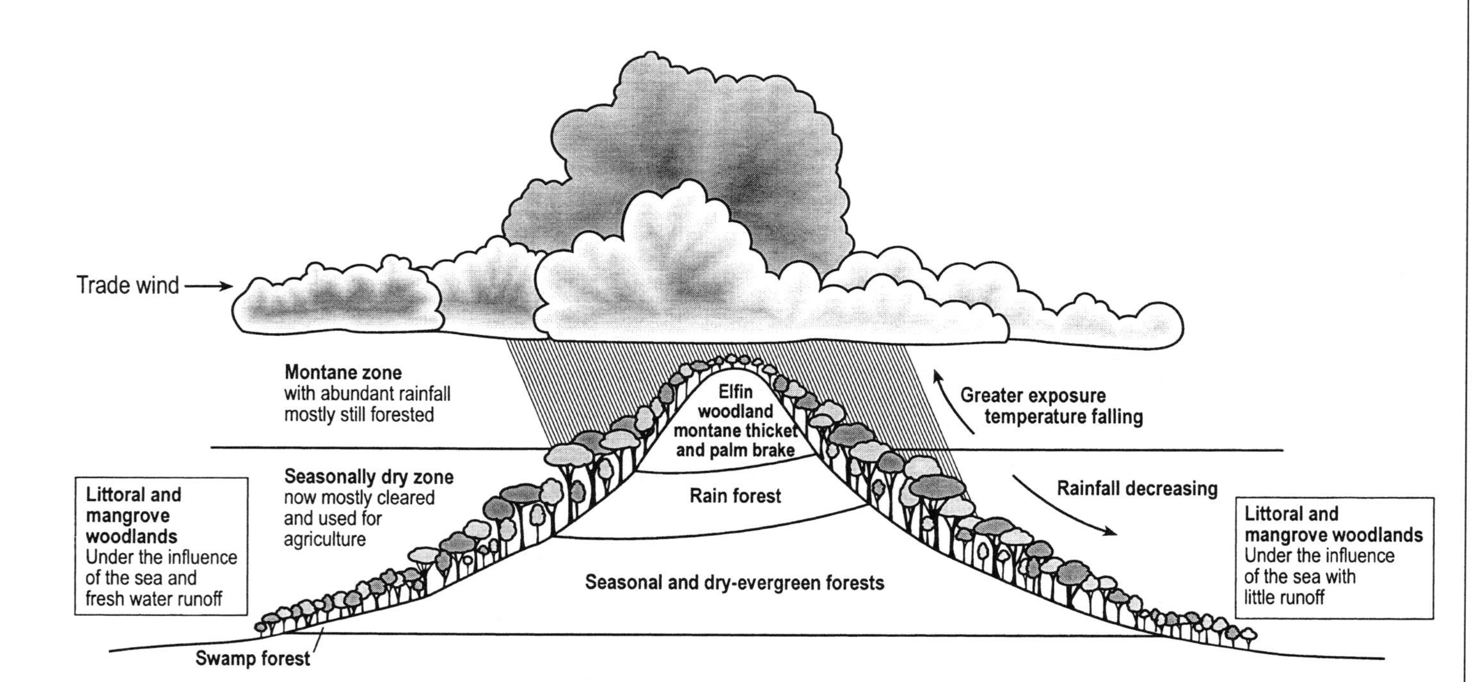

Fig. 3.3. Influence of wind direction on elevation of forest types on small tropical islands. (Adapted from Jordan 1985, with permission of John Wiley and Sons Ltd., publisher)

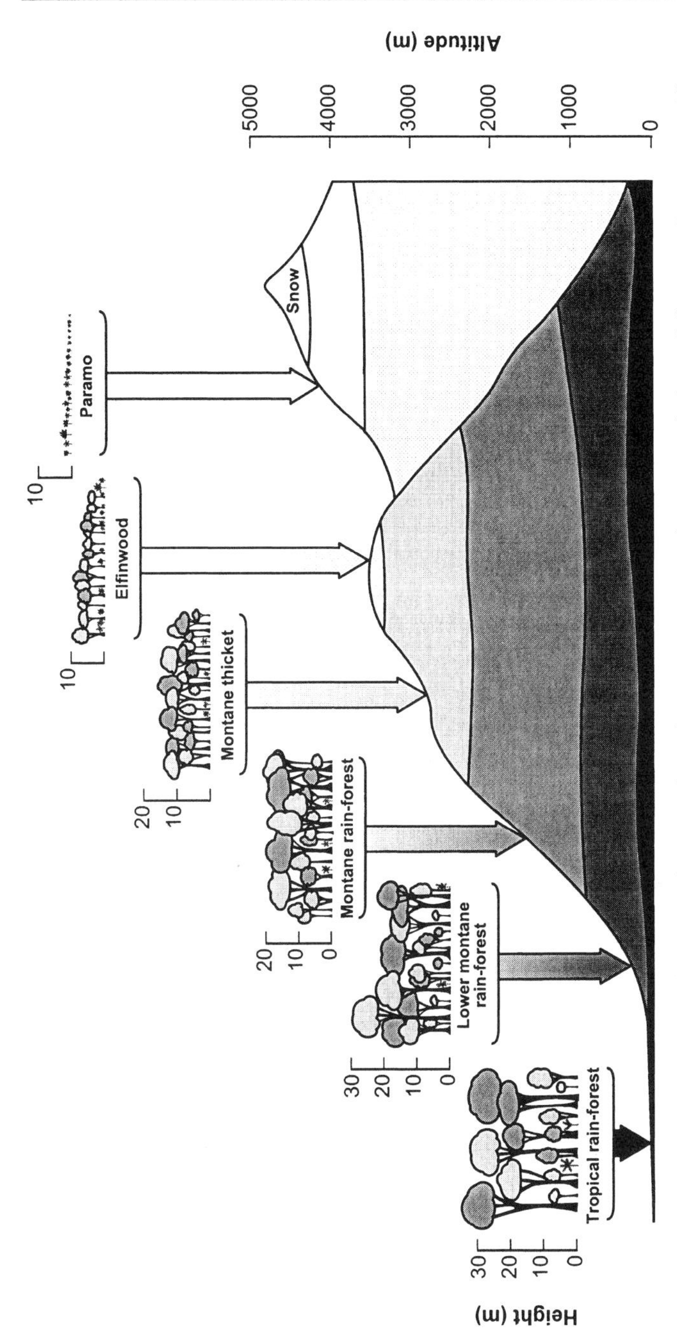

Fig. 3.4. Structure of tropical vegetation along an elevational gradient in continental regions. (Adapted from Jordan 1985, with permission of John Wiley and Sons Ltd., publisher)

with no evapotranspiration). For each combination of region and province, there is a characteristic vegetation type (Fig. 3.2).

Walter (1971) also used climate to delineate global vegetation units. Within the tropics, he recognized equatorial rain forest, tropical rain forests with orographic rains, tropical savannas, rain-green tropical deciduous forests, tropical thorn-steppes, and thorn forests. Figure 3.1 illustrates how forest structure is influenced by climate in lowland regions such as those of Africa, where vegetation formations are finely tuned to the moisture gradient. Beard (1949) classified forests on mountains of small tropical islands according to rainfall and elevation. Elevation of each forest type varies, depending upon the size of the island and the direction of wind relative to the mountains (Fig. 3.3). In continental regions, clouds and rain form at higher elevations than on islands and, as a consequence, the range of each forest type occurs at a higher elevation and covers a greater range (Fig. 3.4; see also Chap. 2).

3.2.1.1 Functional Variation Along Climatic Gradients

Functional characteristics of a tropical forest can be implied, to a certain extent, from their climatic classifications:

- *Energy flow along environmental gradients.* Energy potentially available to tropical ecosystems increases along a gradient from moist to dry forest because of the decreasing cloud cover. However, energy that can be used by plants decreases because moisture scarcity in dry forests limits photosynthetic activity. Therefore, primary productivity decreases along this gradient. Decomposition also decreases because moisture scarcity limits bacterial activity. Energy available to tropical ecosystems, and thus primary production, decreases along an altitudinal gradient from low to high elevations because of increasing cloud cover near the top of mountains. Another factor that limits primary production at high elevations is temperature that is below optimum levels for photosynthetic activity.
- *Nutrient cycling along environmental gradients.* The tightness of nutrient cycles (the efficiency with which nutrients are recycled) will generally decrease along a gradient from wet to dry. In dry forest ecosystems, there are lower rates of decomposition and less potential leaching. Nutrient conservation is less essential for survival of the forest, so fewer nutrient-conserving mechanisms (Chap. 2) are present. Along an elevational gradient, the tightness may increase. Many cloud forests have characteristics such as sclerophyllous leaves (Grubb 1977) and aboveground root concentration (Weaver 1995), characteristics that are similar to forests on nutrient-poor soils (Table 3.1). In cloud forests, nutrient scarcity may occur because of slow decomposition of the litter and humus on top of the soils.

- *Species diversity along environmental gradients.* Species diversity is generally higher in moist forests than in dry forests. It is low, however, where soils are saturated such as in swamp forests or mangroves. Diversity also generally decreases with increasing elevation on mountainsides. For purposes of species conservation, diversity alone is an inadequate index. In high elevation environments, diversity may be low, but there may exist endemic species that are rare or endangered, and whose presence would not be indicated by a simple diversity index.

Although some aspects of forest function are reflected in climatic classifications, others such as tightness of nutrient cycling are not. Because of the prevalence of poor soils in many regions of the lowland tropics and the consequent frequent occurrence of forests adapted to nutrient conservation, climatic classifications alone are often an inadequate system on which to base forest management plans.

3.2.2 Classification Based on Species

3.2.2.1 Classification at the Community Level

The community type found at any given location depends not only on the local climate, but also on the physical and chemical properties of the soil, the topography and elevation, and previous site history. Species integrate all factors of the environment. The functions of all the species in a community combine to produce the function of the community. Therefore, classification based upon communities implies classification based on function. Communities frequently are named after the species that predominate. Richards (1952) described forest communities on the northern edge of the British Guiana forest as follows: "The five communities, in the order they are usually met with from the creek to the ridges, are the *Mora* consociation (*Mora excelsa*), the Morabukea consociation (dominant *Mora gonggrijpii*), the mixed forest association (many dominants), the Greenheart consociation (dominant *Ocotea rodiaei*) and the wallaba consociation (dominant *Eperua falcata*)." Communities along a soil and hydrologic gradient in the Rio Negro region of Venezuela have been given local names: igapó, caatinga, bana, and tierra firme (Fig. 3.5).

Classification based upon community or dominant species is useful in designing management strategies because managers are usually interested primarily in species, since markets demand trees of a given species. Community classification is useful too because such classifications reflect function. For

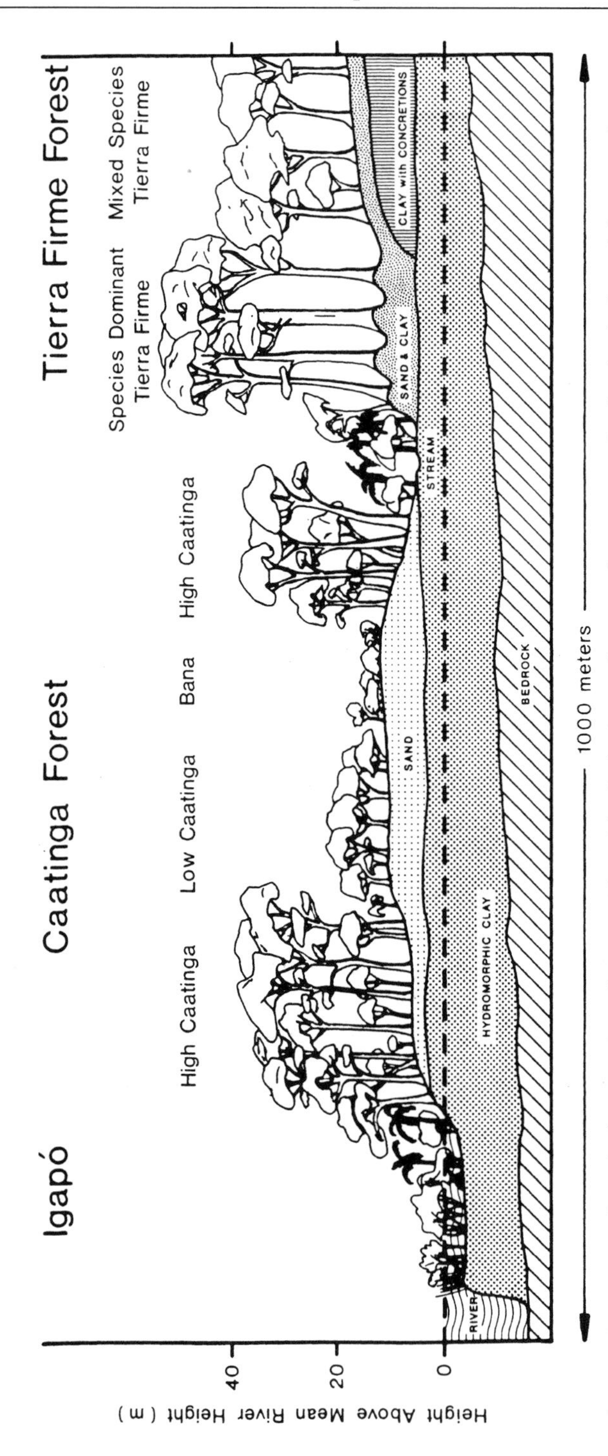

Fig. 3.5. Structure of tropical vegetation along a soil and hydrologic gradient in the Amazon lowlands. (Adapted from Jordan 1985, with permission of John Wiley and Sons Ltd., publisher)

example, the caatinga and bana soils (Fig. 3.5) are low in nitrogen and thus nitrogen-conserving functions, such as production of sclerophyllous leaves, are common. The tierra firme soils are very low in available phosphorus, and thus phosphorus-mobilizing functions, such as secretion of organic acids from roots, may be common (Cuevas 2001).

The changes in species function along a gradient of soil fertility are reflected in a change in species diversity. Diversity is lowest in the bana, intermediate in the caatinga, and higher in the tierra firme. As on mountain slopes and moisture gradients, diversity reflects the environmental stress impinging on the community, except here, soil quality rather than climate governs the level of stress. Communities undergoing the greatest stress due to soil conditions have the lowest potential productivity (Cuevas 2001).

3.2.2.2 Classification Based on "Temperament" of Species

Classification based on the growth and reproductive pattern of individual species is also useful in forest management. Oldeman and van Dijk (1991) described two contrasting growth patterns or "temperament" (sic) of tropical trees that occur within the closed forest. One type of species is called "gamblers", the other "strugglers." Species with a gambler strategy, such as *Didymopanax morototoni* in French Guyana, produce large numbers of seedlings. These cannot survive in the shaded understory but need a light gap in which they can grow rapidly. Light gaps are openings in the forest caused by the death of large canopy trees. The reproductive efforts of gamblers are high in order to increase the likelihood that at least one individual juvenile will find a gap in which to grow.

Species with a "struggler" strategy, such as *Casearia bracteifa*, a small understory species found in French Guyana, produce small numbers of very persistent seedlings. These seedlings struggle but survive, growing very slowly within the densely shaded understory and sometimes even completing their life cycle there. The reproductive efforts of strugglers may be much lower because the life expectancy of each individual juvenile is comparatively high. Oldeman and van Dijk go on to further subdivide. For example, there are "strugglers in extreme", which are capable of completing their life cycle within the shade. Other strugglers produce juveniles that are shade-tolerant but need light gaps to mature and grow into the canopy. This type of physiological understanding about the functional role of particular species is necessary for management of that species. For example, shelterwood forestry (discussed in Chap. 5) is particularly suitable for species that are shade-tolerant but respond to opening up of a light gap. The creation of a gap by the harvest of an individual tree mimics the naturally occurring gaps.

3.2.2.3
Classification Based on Successional Stage

Communities that invade recently abandoned land or pasture, or other disturbances that destroy the structure of the mature forest, are called successional forests. Budowski (1965) listed the characteristics of early, middle, and late successional species (Table 3.2). Early successional species are often characterized as fast growing. Although secondary forests grow fast in volume, most pioneer tree species have wood of low specific gravity, sometimes 0.3 or less (Jordan and Farnworth 1980). This makes them more valuable for uses such as pulp. Species with fast volume increment also are desirable for land

Table 3.2. Characteristics of four successional stages of tropical forest communities. (Adapted from Budowski 1965)

Characteristic	Successional stage			
	Pioneer	Early secondary	Late secondary	Climax
Age (years)	1–3	5–15	20–50	>100
Height (m)	5–8	12–20	20–30	30–60
No. of tree species	1–5	1–10	30–60	100+
Distribution of dominants	Very wide	Very wide	Wide	Endemics common
Growth (volume)	Very fast	Very fast	Mixed	Slow
Life span of dominants	Less than 10 years	10–25 years	40–100 years	>100
Shade tolerance of seedlings	Intolerant	Intolerant	Tolerant	Tolerant
Regeneration of dominants	Scarce	Absent	Absent or large mortality	Abundant
Dissemination of seeds	Birds, bats, wind	Wind, birds, bats	Mainly wind	Gravity, mammals
Specific gravity of wood	Light	Light	Medium	Heavy
Diameter of stems	Small	<60 cm	Some large	Many large
Size of seed	Small	Small	Medium	Large
Viability of seeds	Long	Long	Medium	Short
Shrubs	Many but few species	Abundant but few species	Few	Few but many species
Grasses	Abundant	Abundant or scarce	Scarce	Scarce

reclamation because they will more quickly form a canopy that protects the soil from erosion.

From a forestry perspective, the functional characteristics of early successional species or pioneer species make them relatively easy to manage (see Chap. 5). Clear-cutting may be necessary to obtain the necessary light for germination and growth of desired species. The pulse of nutrients released by burning or other forest disturbance may also stimulate the growth of early successional species, such as species of the genera *Cecropia*, *Musanga*, *Macaranga*, *Ochroma*, and *Trema*. These species produce large quantities of small seeds that are capable of dormancy and are called "orthodox" (Whitmore 1998). Orthodox seeds usually remain viable for relatively long periods of time and form the seed bank in many forests. Dormant seeds are often already present in the cleared area and more are quickly blown or carried in after an area is cleared.

In contrast to management for pioneer species, management for species of the mature forest is often difficult due to seed characteristics. Seeds of many primary forest species are large and heavy and fall straight down. Other species, such as those of the commercially important Dipterocarpaceae family, have winged seeds, but dispersal is still not far from the parent tree. Other characteristics relate to the viability of seeds. Almost all the seeds of some species are viable, while for other species only a few germinate. Length of time that seeds remain viable also varies. Seeds of some mature forest tree species often have little or no dormancy. They germinate as soon as they reach the forest floor, or within a few days. Unless they encounter favorable conditions within days after falling, they never germinate. These types of seeds are called "recalcitrant" (Whitmore 1998), somewhat of a misnomer according to Janzen and Vázquez-Yanes (1991) who said, "Such a terminology is akin to labeling an animal 'recalcitrant' if it cannot be maintained without food."

The characteristics of many mature forest species mean that most will spread only very slowly unless the seedlings are transported and planted by humans. It also means that the seeds must be collected very soon after falling if they are to be useful. Specific procedures to preserve, store, and germinate the seeds when desired must be developed for each individual species. Because of such special requirements, the physiological ecology of each species must be well understood in order to achieve successful management. The high specific gravity of the wood of many mature forest species is also a consideration in forest management because it means that the volume accumulation of these species will be relatively slow. However, such high-density wood often brings a good price. It is also often resistant to decay and can be used for pilings and other submerged support beams.

The secondary successional progression outlined in Table 3.2 is often called relay floristics (Whitmore 1991) and frequently occurs on large areas of abandoned agricultural land when there is no further disturbance. Early pioneers can modify the environment in such a way as to make establishment of later species more feasible. Such modifications include nitrogen enrichment of the soil by nitrogen-fixing bacteria associated with leguminous tree species such as those of the genera *Inga*, *Acacia*, and *Albizia*. Other influences of pioneer species on the microenvironment may include provision of shade which increases soil moisture and lowers soil temperature (necessary for many mature forest species). Pioneer species can also provide habitat for seed-dispersing animals. However, if the soil in the successional area has not been heavily disturbed, microhabitat may be more favorable for all species, and the successional progression may resemble more simultaneous colonization in which the species that become established depend simply upon which seeds arrive first in the disturbed area.

When secondary succession progresses according to the pattern of relay floristics (Table 3.2), we can expect to observe the following ecological trends which are potentially important for management:

- Species diversity and mutualisms will increase.
- Gross photosynthesis increases at first, as the quantity of leaves in the successional forest increases. Respiration also increases, but not as quickly. Net primary production (the difference between gross photosynthesis and respiration) increases in early stages of succession and as a result accumulated biomass increases. Later, however, more and more of the gross photosynthesis is used for maintenance and less for biomass accumulation. Eventually, biomass of the forest reaches an approximate steady state (Fig. 2.7).
- The trend of nutrient cycling during the course of succession may depend on the nature of the soil on which the succession is occurring. If the soil is low in nutrients, the cycle may become tighter as nutrient-conserving mechanisms play an increasingly important role (Odum 1969). If the soil is high in nutrients, the cycle may become looser, because early in succession, available nutrients go towards building biomass. However, as maturity approaches and biomass no longer increases, available nutrients are leached from the soil (Vitousek and Reiners 1975).

3.2.3 Forest Classification Based Upon Soil Nutrient Status

A forest classification that considers nutrient status could be made according to soil fertility. For example, in the humid Amazon lowlands, Spodosols, Oxisols, Ultisols, and Alfisols represent a gradient of increasing soil fertility. However, to classify a forest as "forest growing on an Oxisol" emphasizes characteristics of the soil instead of the forest. To characterize tropical forests according to their degree of adaptation to low soil nutrients, Jordan (1985) classified forests along a continuum from those adapted to extremely low nutrient supplies (oligotrophic forests) to those on relatively fertile soils (eutrophic forests). The characteristics used in this classification are listed as parameters in Table 3.1.

Net primary productivity increases along the gradient from oligotrophic to eutrophic forests. In addition, species diversity may increase along the gradient. For example, the high caatinga forest on Podsol sands of the Amazon is a highly oligotrophic forest. Compared to the eutrophic forest in Panama, it has a high root biomass, a slightly lower above-ground biomass, a high root:shoot ratio, a high concentration of roots above the soil surface, a low specific leaf area (the leaves are thick and tough), a low leaf area index, an average leaf biomass (the thickness of the leaves compensate for the lower number of leaves), a slow rate of leaf decomposition, and low concentrations of nutrients in the leaves.

Nutrient use efficiency (the amount of biomass produced per unit of nutrient taken up by plants) is an indication of the degree to which a nutrient is limiting (Vitousek 1982, 1984). Because nutrient use efficiency is an indicator of the tightness of the nutrient cycle, these values are used in parameters 12 and 13 of Table 3.1. Nitrogen use efficiency is highest in the caatinga forest, but phosphorus use efficiency is highest in the Oxisol forest. Estimation of values of nutrient use efficiency for a number of sites at high elevations and in lowland forest in the tropics has led to the hypothesis that phosphorus is frequently a limiting nutrient in the lowlands, while nitrogen is frequently limiting at high elevations (Vitousek 1982, 1984). Phosphorus is limiting in the highly weathered soils of the lowlands because it is immobilized by the iron and aluminum, while nitrogen is limiting at high elevations because it is immobilized in litter that decomposes very slowly due to low temperatures (Jordan 1985).

3.2.3.1 Implications for Management

- Species diversity is greater on relatively rich soils than on poorer soils. However, species diversity does not indicate the abundance of endemic species, which may be of interest to conservationists. If a desirable species is rare, it does not matter if it occurs on nutrient-rich or nutrient-poor sites. In either case, care must be taken to ensure its reproductive success.
- Productivity and decomposition are lower on nutrient-poor soils, because the metabolism of trees and decomposers is limited by a scarcity of nu-

trient elements. Because oligotrophic forests are less productive, their economic potential is less. They might serve better as nature reserves than as sources of timber.

- Nutrient cycles will be tighter on nutrient-poor soils than on nutrient-rich soils. If management is carried out on nutrient-poor soils, care must be taken to preserve the soil organic matter and humus and litter on top of the soil. Otherwise, the productive potential of the site will quickly deteriorate.

3.2.3.2 The UNESCO Classification System

The UNESCO classification system (Table 3.3) is a comprehensive classification based on vegetation structure and environmental variables that reflect both stress and function (climate, soils, topography, successional state). However, the vegetation formations determined with this system are too large to be useful in predicting management requirements for a particular stand. Where good forest community maps are available, they can be incorporated into the UNESCO system to serve as a guide to forest management.

The map of forest communities in the Luquillo Experimental Forest of Puerto Rico (Fig. 3.6) was developed using the UNESCO system. The Tabonuco forest dominated by *Dacryodes excelsa* corresponds to the submontane to montane forest in the physiographic climax formation of UNESCO. Tabonuco is a valuable timber species that has good form and its response to management is well known (USDA 1990). Above Tabonuco on the mountain slope is the Colorado community, dominated by *Cyrilla racemiflora*. This community, which corresponds to UNESCO's alto-montane forest (moist) category, is much less valuable for timber production but has been used for charcoal in the past. It is valuable as a habitat for wildlife, most notably the endangered Puerto Rican parrot. On steeper slopes and wetter soils, the forest is dominated by the sierra palm, *Prestoea montana*, which has little commercial value. On the highest peaks of the mountain the dwarf forest occurs, composed of dense stands of short, small-diameter, twisted trees and shrubs heavily covered with mosses and epiphytes. This would correspond to UNESCO's alto-montane moss forest (wet, misty).

A major problem with the UNESCO system is that it must be used in conjunction with accurate maps of forest communities. Due to the time and expense of preparing them over large areas such maps often are not available.

Table 3.3. The UNESCO classification system of tropical vegetation formations. The table presents a climatic gradient from the climatic equator towards higher latitudes, and the corresponding gradients of plant habitus and of the main plant formations. Habitus is expressed as the type of climatic adaptation and corresponding morphological adaptation. Formation refers to formation group and formation levels of ecological–structural classification. Climatic data refer to tropical lowlands below 300 m altitude. (UNESCO 1978)

	Annual insolation on the ground (kcal/cm^2)					
	SE Asia 140–160, Congo 120–130, and the Amazon 100–120 kcal/cm^2	ca. 160	ca. 180	ca. 200	ca. 220	>220
Mean annual temperature (°C)	28	25	21–32	20–33	Extreme variation	Extreme variation
Annual variation (°C)	3	15–20	30	35		
Diurnal variation (°C)	9	20	20	30		
Wind	Predominance of tropical low-pressure trough, low velocities except in convectional squalls and local tornadoes	High velocities during summer (typhoon, hurricane, cyclone), low during dry season, strong effect of tropical convergence zone. Local frontal storms toward end of dry season	Seasonally dry tropic or moist tropic air (trades, monsoon), velocities moderate except during passage of cyclones	Predominance of tropic high-pressure cell, average wind speeds low to moderate, high velocities in advective storms	As before, dry and hot storms more frequent	As before, dust storms common
Relative humidity	95/100% night, 60/70% day, little seasonal variation	90/100% wet season, 60/80% dry season	80/90% wet season, 40/60% dry season	60–80% wet season, 20–50% dry season	Usually 50%	Average very low, but locally very high for short periods

Table 3.3 (continued)

	Annual insolation on the ground (kcal/cm^2)					
	SE Asia 140–160, Congo 120–130, and the Amazon 100–120 kcal/cm^2	ca. 160	ca. 180	ca. 200	ca. 220	>220
Ratio of potential evaporation to precipitation: Ep/Po	<1	<1	>1	>2	>1	>8
Precipitation (mm/year)	2,000	1,300–3,000	500–1,500	350–1,000	100	<100
Distribution of precipitation	Even, <2 dry months	2 or 4 dry seasons, 3–5 dry months	2 or 4 seasons, 6–8 dry months	2 seasons, 9–10 dry months	11 dry months	12 dry months
Location	Equatorial belt and areas with constant moist air masses outside this belt	Subequatorial to outer tropics with influence of trades, monsoons, and monsoon-like alternating winds	Subequatorial to subtropical summer-rain belt	Outer tropical belt, summer rain belt	Outer tropical to subtropical belt of descending air masses	Outer tropical to subtropical belt of descending air masses
Climate type	Tropical perhumid (wet), non-seasonal, diurnal variation >annual variation	Tropical humid (moist), isotherm, seasonal with predominantly summer rainfall	Tropical dry seasonal with summer rainfall	Tropical very dry, seasonal with summer rainfall	Tropical arid	Tropical arid
Growing period (months)	11–12	7–10	4–6	2–3	1–2	
Characteristic habitus of climatic climax formation	Megatherm-hydrophilous, hygromorph-mesomorph	Megatherm-tropophytic, tropomorph	Megatherm-tropophytic	Megatherm-tropophytic, xeromorph	Sclerophytic-xerophytic	Bare

Table 3.3 (continued)

	Annual insolation on the ground (kcal/cm^2)					
	SE Asia 140–160, Congo 120–130, and the Amazon 100–120 kcal/cm^2	ca. 160	ca. 180	ca. 200	ca. 220	>220
Main climatic climax formations	Superhumid to humid ombrophilous, evergreen tropical forest and semi-evergreen wet forest	Humid to sub-humid, semi-deciduous and deciduous tropical forest	Sub-humid to semi-arid deciduous tropical forest	Semi-arid deciduous tropical thorn woodland	Arid thorn scrub and semi-desert	Perarid desert
Edaphic climax formations	Littoral forest; mangrove forest and woodland; freshwater swamp forest and grassland; peat swamp forest; riparian forest; single-dominant forests on certain soils; sclerophyll forest	Littoral forest; mangrove forest and woodland (less luxuriant); freshwater swamp forest and grassland; riparian forest (often relic gallery); evergreen forest (on moister well-drained soils); sclerophyll forest (sandy terraces, skeletal soils)	Littoral woodland; mangrove forest to scrub; riparian fringing forest and grassland; sclerophytic thorn woodland; deciduous moist forest	Littoral scrub; mangrove woodland to scrub; riparian fringing woodland and grassland; xerophytic semi-desert scrub	Littoral scrub; mangrove scrub; riparian fringing; scrub and grassland	As in the arid zone, but rarer and poorer

Table 3.3 (continued)

	Annual insolation on the ground (kcal/cm²)					
	SE Asia 140–160, Congo 120–130, and the Amazon 100–120 kcal/cm²	ca. 160	ca. 180	ca. 200	ca. 220	>220
Physiographic climax formations	Submontane forest (locally rich in oaks and laurels, simpler structure than the climatic climax); montane forest (moist); alto-montane forest (wet, misty); alto-montane woodland and scrub (moist)	Similar to the perhumid zone, except for species composition, conifers increase in southern and northern hemispheres, bamboo species become more frequent in the northern hemisphere	Similar to the humid zone, but relative effect of exposition and barrier-effect more pronounced	Similar to dry zone, very strong and noticeable effect of elevation, frost occurs regularly even at lower altitudes, particularly in hollows. Creates local dwarf vegetation. Generally more open and scrub-like. Barrier effect very pronounced	As before	As before
Degraded formations	Subseral secondary forest; disclimax secondary forest; disclimax pine forest to pine savanna (mostly at higher altitude); disclimax grassland (*Imperata cylindrica* usually common); disclimax karst woodland; disclimax sclerophytic savanna	Subseral secondary forest; disclimax secondary forest; disclimax savanna; disclimax pine forest, pine woodland or pine savanna (higher altitudes); disclimax karst woodland; disclimax sclerophytic or xerophytic savanna	Disclimax xerophytic savanna; disclimax thorn scrub; disclimax semi-desert shrub	Disclimax xerophytic thorn scrub; disclimax semi-desert	Disclimax desert	Bare

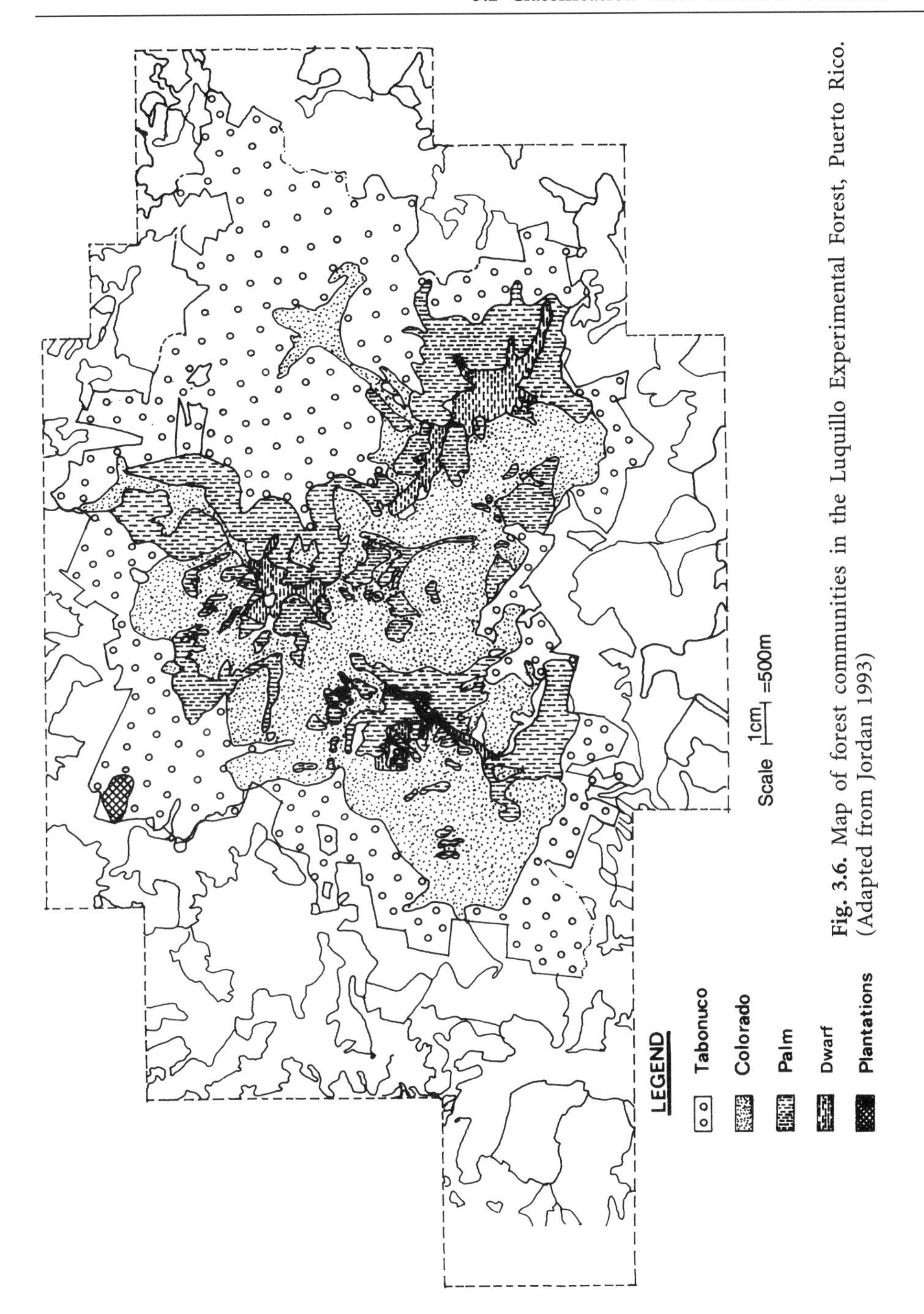

Fig. 3.6. Map of forest communities in the Luquillo Experimental Forest, Puerto Rico. (Adapted from Jordan 1993)

3.3 Conclusion

There appears to be no single classification scheme that is general enough to be of use to everybody yet particular enough to take local conditions into account in a way that is meaningful for management. Correlations between species and forest function based on local knowledge still are the most useful basis for small-scale forest management. Remote sensing techniques along with geographic information systems offer the capability of defining forest communities on a scale useful for management over larger areas.

Deforestation in the Tropics

4

4.1
Rates of Deforestation

Tropical deforestation was a concern to colonial powers in the 19th century, because of their reliance on the forests of their colonies to supply timber for building naval vessels. For example, British Colonial Foresters became concerned with the lack of regeneration of teak forests in Burma (now Myanmar) and initiated a reforestation program in the mid-1800s (Takeda 1992). However, global concern over tropical deforestation did not begin until the mid-20th century. Following World War II, logging increased in intensity and scale, and population pressures in tropical countries resulted in the clearing of tropical forests for agriculture. By the 1960s, scientists began to recognize that the disappearance of tropical forests represented an important loss of global resources, and in the 1970s and 1980s began to determine the area of tropical forests worldwide and the rate at which they were disappearing (Table 4.1).

The estimates varied widely due to a number of factors. First, boundaries of forests are not always well delineated. Along a climatic gradient, density of trees may gradually decrease. For example, a gradient could begin with a forest with a closed canopy that transitions to a savanna with open spaces between trees; the savanna gives way to a grassland or shrubland. Different

Table 4.1. Estimates of the global area of tropical moist forest, and rate of deforestation from the 1970s to 1990. Values for area are in millions of hectares. Rate of deforestation is in millions of hectares per year. (Adapted from Grainger 1993; Pearce and Brown 1994; Achard et al. 2002)

Period	Area of forest	Area deforested/annum
1970s	935–972	7.5–20
1976–1980	1,081	6.1
Late 1980s	800	14.2
1981–1990	1,282	15.4–16.8

authors may delimit each forest type differently. Most studies have limited their estimates to "moist" forests, probably upland moist forests (Table 4.1). Other types of moist forest such as swamps, given in Table 3.3 (Chap. 3), would not be included if some of the data in Table 4.1 were limited to upland moist forests. Exclusion of open forests or dry forests results in further discrepancies. The pre-1990 estimates for forest cover in Table 4.1 are low compared to the estimates in Table 4.2, because the latter includes dry deciduous forests (forests with greater than 4 months' dry season and that lose all their leaves during that time).

Another problem with data on extent of forest cover and rates of deforestation is that, in many cases, values for individual countries are obtained from documents prepared by local authorities who often rely on maps drawn by field workers with little training. The authorities themselves may have political motivations for either exaggerating or under-reporting the extent of forest cover. Often maps are not available for all tropical countries or all regions within a country and estimates may be based upon extrapolations from available sources.

The more recent data of Achard et al. (2002; Table 4.3) were derived on a global basis from satellite imagery, but with varying degrees of resolution.

Table 4.2. Estimate of areas and rates of deforestation of all tropical closed forest, all open tropical woodlands, and the sum of the two, in 1980. Values for area are in millions of hectares. Rate of deforestation is in millions of hectares per year. (Adapted from Grainger 1993)

	Latin America		Africa		Southeast Asia		Total	
	Area	Rate	Area	Rate	Area	Rate	Area	Rate
All closed forests	679	4.1	217	1.3	306	1.8	1201	7.3
Open woodland	217	1.3	486	2.3	31	0.2	734	3.8
Closed plus open	896	5.4	703	3.6	337	2.0	1935	11.1

Table 4.3. Estimate of area of tropical moist forest by continent, and rate of deforestation between 1990 and 1997. Values for area are in millions of hectares. Rate of deforestation is in millions of hectares per year. (Adapted from Achard et al. 2002)

	Latin America	Africa	Southeast Asia
Forest cover, 1990	669	198	283
Forest cover, 1997	653	193	270
Area deforested/annum	2.5	0.85	2.5
Annual percent loss	0.38	0.43	0.91
Area of regrowth/annum	0.28	0.14	0.53

Fearnside and Laurance (2003) have criticized the Achard et al. (2002) data on the basis that it did not include dry forests; estimates on biomass were based on data from a single site; it failed to include palms, vines, and understory vegetation; and erroneously assumed that secondary forests would regenerate 70% of their biomass in 25 years. This may account for the relatively low rate of deforestation reported in Table 4.3.

The World Resources Institute (2004) presents data from the US Geological Survey's Earth Resources Observation System on forest coverage in 1993 (Table 4.4). In their data, evergreen broadleaf forests are those that have a percent canopy cover greater than 60%, and almost all trees remain green throughout the year. Deciduous forests have canopy cover greater than 60% with annual cycles of leaf fall. Estimates diverge somewhat from those in Tables 4.2 and 4.3, but differences may be due to different systems of classification. The largest discrepancies are for Africa. The evergreen and deciduous forests for Africa shown in Table 4.4 do not include the open woodland shown in Table 4.2.

A further problem is simply defining what constitutes "deforestation". Does a trail made by loggers to extract one mahogany tree constitute deforestation? Does the clearing made by one shifting cultivator who enters the forest by that trail qualify? When loggers remove many of the trees, but leave some individuals still standing, is the area deforested? As the peasant farmers stream in and lay claim to the land, eventually it does become "deforested". What percentage of cover removal constitutes "deforestation?" Determining the point at which closed forest becomes deforested is quite arbitrary.

Further complicating the problem is the fact that there are many areas, as in the eastern Amazon region, where deforested land has been abandoned and is growing back into secondary forests (Dubois 1990). These stands are comprised of species different from those in closed, primary forests. Should the areas of these recovering forests be subtracted from the areas of primary forest that is cleared? Plantation forests growing on previously deforested land present the same question.

Laurance et al. (1997) pointed out that there is an increase in biomass loss in forest fragments, due to the fact that along fragment edges, microclimatic

Table 4.4. Tropical forest cover in 1993 characterized from 1-km advanced very high resolution radiometer data (USGS 2004). Values are in millions of hectares

	Latin America	Sub-Saharan Africa	Southeast Asia
Evergreen broadleaf	784	350	200
Deciduous	47	14	–
Total	831	364	–

changes and elevated wind turbulence cause increased tree mortality. Such edge effects increase exponentially as the area cleared increases linearly. Losses due to edge effect would not be detected by satellite reconnaissance, and probably not even by conventional land-based mapping techniques. Laurance and Fearnside (1999) further suggest that losses may be increasing. They stated that "Despite initiatives to reduce deforestation, in 1998 the rate of forest loss in the Brazilian Amazon rose by nearly 30% over the preceding year – not including the extensive areas degraded by ground-fires, logging or habitat fragmentation". Consequently, rates of deforestation may be greater than those reported in Tables 4.1–4.3.

Despite the problems of assigning accurate numbers to deforestation rates and biomass changes, the data suggest regional trends. From 1990–1997 Latin America and Southeast Asia had similar rates of deforestation in terms of area cleared. However, because in Latin America there is a much greater area of moist forest, percentage loss is lower (Table 4.3). The rate of loss in open woodlands of Africa is relatively high, but Africa has considerably more open woodlands than closed forests, and more open woodlands than the other two continents (Table 4.2).

4.2 Causes of Deforestation

4.2.1 Proximate Causes of Deforestation

4.2.1.1 Expansion of Agriculture

Myers (1984) cited "shifting cultivation" as the most important cause of deforestation. There are various types of shifting cultivation. A destructive type of shifting cultivation is practiced by non-indigenous colonizers who often know little about farming, other than chopping down the forest, burning it, and planting corn or rice in the ashes. After 2 or 3 years, production gives out, and they are forced to move further into the wilderness. They may just abandon the land, or they may sell their land to a consolidator, perhaps a rancher, who is buying up land in the region, either for pasture or for speculation. In Brazil, some of these shifting cultivators are from the drought-ridden north-east and have migrated into the Amazon rain forest and cleared small patches for agriculture. Although they may follow roads built by the government or by loggers, they act on their own. Other colonizers may be participants in a government resettlement program, in which people from other regions of the country, including cities, are transported to wilderness areas and given some land and subsidized housing. Myers (1992) referred to this type of shifting cultivation as "*shifted* cultivation", i.e. practiced by people who would not be farmers, given the choice, in contrast to "*shifting* cultivation", practiced in a more "traditional" fashion. In the 1980s and 1990s, "shifted" cultivation apparently accounted for 35, 70, and 50% of deforestation of closed forests in America, Africa, and Asia, respectively.

Cultivation of illegal narcotic plants such as coca (the source of cocaine) and opium (the source of heroin) in rain forest clearings in Southeast Asia and South America is also a destructive type of shifting cultivation. When the plots are discovered by drug enforcement agents, the farmers move elsewhere.

Another type of shifting cultivation has been practiced by indigenous peoples for subsistence and by people whose ancestors moved into the forest and learned traditional techniques. A small area, approximately 1 ha, is cleared and then burned. Both annual crops such as manioc and perennial crops such as fruit and nut trees are planted. The annual crops produce well for 1 or 2 years, but declining nutrient availability and weed pressure rapidly diminish the production by annuals. However, by the third year, perennial crops such as plantain and cashew have become established and begin to yield. Successional species that are valuable for wood, medicines or other uses are favored, and other species may be weeded out.

In some areas, people who have lived in the forest for generations follow this same system. This system can be sustainable, as long as population density is low, because the land can be left fallow long enough to recover soil fertility. Organic debris from the surrounding forest quickly covers the soil, either directly through leaf and litter fall or through dispersion by animals. It is the organic matter that keeps the nutrients available in most tropical soils. When the plot is left fallow, the soil organic matter gradually builds up again, and the nutrients, especially phosphorus, become available (Jordan 1995 a). Some tribes, like the Kayapó in Brazil, plant perennial crops and fruit trees in the fallow, managing the plot as a long-term rotation (Posey 1982).

The length of time required for the site to regenerate sufficient nutrients to permit further cultivation depends on the soil quality and the intensity of cultivation, and can vary from a few years to almost a century (see also Chaps. 2 and 5). Younger, volcanically derived soils, for example, regenerate more quickly than highly weathered Oxisols of the lowland tropics. As forest areas become more populated, the fallow period becomes shortened. Because of the short fallow, nutrient stocks in the soil do not fully recover, and the period of cultivation must be shortened (Nye and Greenland 1960).

Permanent cultivation has become a more important cause of deforestation than shifting cultivation (Geist and Lambin 2002). In South and Central America, large areas of forest have been converted to pasture. Mega-farms that produce soybeans are encroaching on the southern fringes of the Amazon rain forest (Nepstad et al. 2002). Tree-crop plantations for rubber, oil palm, cocoa, coffee, and coconut have been an important cause of deforestation in Africa and Southeast Asia (Grainger 1993). In Central America, plantation crops such as coffee, cacao, palm oil, bananas, pineapple, and others have also been a major force driving deforestation. Timber is another key plantation crop. In the Jarí project in the state of Pará in Brazil in the eastern Amazon, tens of thousands of hectares of primary forest were cleared in order to plant the fast-growing species *Gmelina arborea* for pulp wood (Jordan 1995b). Commodity crops such as those mentioned above are subject to global economic cycles of boom and bust. When demand peaks, deforestation occurs as land is cleared for new plantings. When oversupply occurs, cleared land is often abandoned.

Tropical agriculture is not necessarily ecologically or economically unsustainable. The colony at Tomé Açu in the state of Pará, Brazil, is an interesting example of colonizers who learned to farm sustainably on the poor soils of the Amazon. Japanese immigrants in the 1920s settled the area and experimentally devised a rotational scheme that included mixtures of perennials and annuals, and various animal stocks to supply manure as well as meat and milk (Subler and Uhl 1990). Because of the diversity of their agricultural practices and products, their soils retain fertility and their economic income is less influenced by global cycles that affect individual commodities.

Ranching

Conversion of tropical forest to pasture has been a particularly important activity in Brazil. In the late 1970s, the Brazilian government instituted a program to encourage ranching in the Amazon region. Large cattle enterprises were promoted as the prototype for development. There was a perception that ranching actually improved the quality of the soil by increasing soil nutrients (Barbosa 2000). However, the soil tests that led to this conclusion presumably were taken immediately after the cutting and burning of the forest, before the ashes containing the nutrients had leached away (see Figs. 2.10–2.12 for nutrient dynamics following burning of forest, and see Fig. 2.13 for changes in the availability of soil nutrients following conversion of forest to pasture in the Amazon region of Brazil). Roughly 10 million ha of land was converted from forest to pasture by the early 1980s (Hecht 1985). Since then, government incentives to clear forest have been reduced, but conversion of forest to pasture still remains important in the Amazon region (Castellanet and Jordan 2002).

4.2.1.2
Wood Extraction

Wood has always been an important fuel for forest dwellers, and even in the mid-20th century 80% of all wood harvested was for fuel (see Chap. 1). Commercial loggers in the tropics concentrated only on a few valuable timber species such as teak and mahogany. Following World War II, woods of lighter density were also extracted to be sold in national or foreign markets. From 1950–1980, tropical hardwood exports rose 14-fold, from 4.6 to 61.2 million m^3 of roundwood per year (FAO 1989; see also Chaps. 1 and 6).

In the 1990s, there was a dramatic increase in logging in the tropics. In many developing nations, tariffs and trade barriers fell, while new international free-trade agreements promoted foreign investment, particularly in natural resource-based industries such as timber (Laurance 2000). Each year, approximately 6 million ha of tropical forest was logged (Whitmore 1997). Although only between 2 and 40 trees may be harvested per hectare, 10–40% of the remaining forest biomass is killed or severely damaged during logging operations (Uhl et al. 1991; Verissimo et al. 1992). The main damage results from the labyrinth of roads, bulldozer trails, and small clearings in the forest. The heavy machinery kills many smaller trees, damages and compacts the soil, and increases soil erosion and stream sedimentation (see also Chap. 5). A study in Indonesia found that even though only 3% of the trees were cut, the logging operation damaged 49% of the trees in the forest (Urquhart et al. 1998). Once opened, the forest is increasingly vulnerable to hunters, ranchers, and shifting cultivators (Wilkie 1992). Logging also increases the vulnerability of forests to fire by rupturing the forest canopy and creating piles of dry, flammable debris. In the Amazon and Borneo, millions of hectares of logged forest were destroyed by wildfires during the 1982–1983 and 1997–1998 El Niño droughts (Brown 1998; Laurance 1998).

4.2.1.3
Development of Infrastructure

The settlement and subsequent clearance of frontier lands in Latin America have closely followed the expansion of the road network. Road building is not always carried out exclusively by governments. In Ecuador, the early penetration roads into the environmentally fragile eastern region were largely built by multinational oil companies. Mahar and Schneider (1994) argue that road building is the single most powerful factor causing the deforestation of frontier areas in Latin America. Certainly oil exploration, agricultural expansion, and timber extraction are not possible without roads, and they are important reasons why roads are built in many forest regions.

The results of providing all-weather overland access to frontier areas are often cumulative and irreversible. The increase in population associated with the completion of primary roads usually generates demand for secondary and feeder roads, which in turn attract more population. The process has been documented throughout the Brazilian Amazon, in the eastern lowlands of Ecuador, Peru, and Bolivia, as well as in Central America. Once roads have been built into wilderness areas, there is pressure on the local and national governments to provide further infrastructure such as health services, education, police, and other social services (Castellanet and Jordan 2002).

4.2.2 Underlying Causes of Deforestation

Agriculture, ranching, timber extraction, road building, and mining are only the proximate causes of deforestation (Geist and Lambin 2002). These activities are necessary for human survival, but individuals cannot easily undertake them on their own. There have to be institutions that make these activities feasible, and forces that drive people and institutions to undertake them. These institutions and forces are the underlying causes that drive tropical deforestation and can be separated into five major categories (Fig. 4.1). They are presented here in order of decreasing importance as listed by Geist and Lambin (2002).

4.2.2.1 Economic

The economic development on forested frontiers often follows a pattern that is exemplified by the region west of Belém, Pará, on the eastern edge of the Amazon rain forest (Jordan 2001; Fig. 4.2). In this figure, the horizontal axis is the distance from a particular piece of land to markets for agricultural products or beef. In the eastern Amazon region, the city of Belém is a traditional market and the town of Paragominas is becoming one. "Distance from markets" not only implies the actual linear distance, but also includes the logistical difficulty of getting supplies from the market and taking the produce to the market. A farm 100 km from the market on a paved road is logistically closer to the market than a farm 40 km away on a dirt road that becomes impassable during rainstorms. "Net present value" of the land is simply its current market price. Net present value is determined in part by the value of the labor necessary for economic income; the cost of defending property rights to that land; the value of the capital necessary to produce an economic income; and the equilibrium price of the land, that is, what bidders in the market place will pay after consideration of other factors.

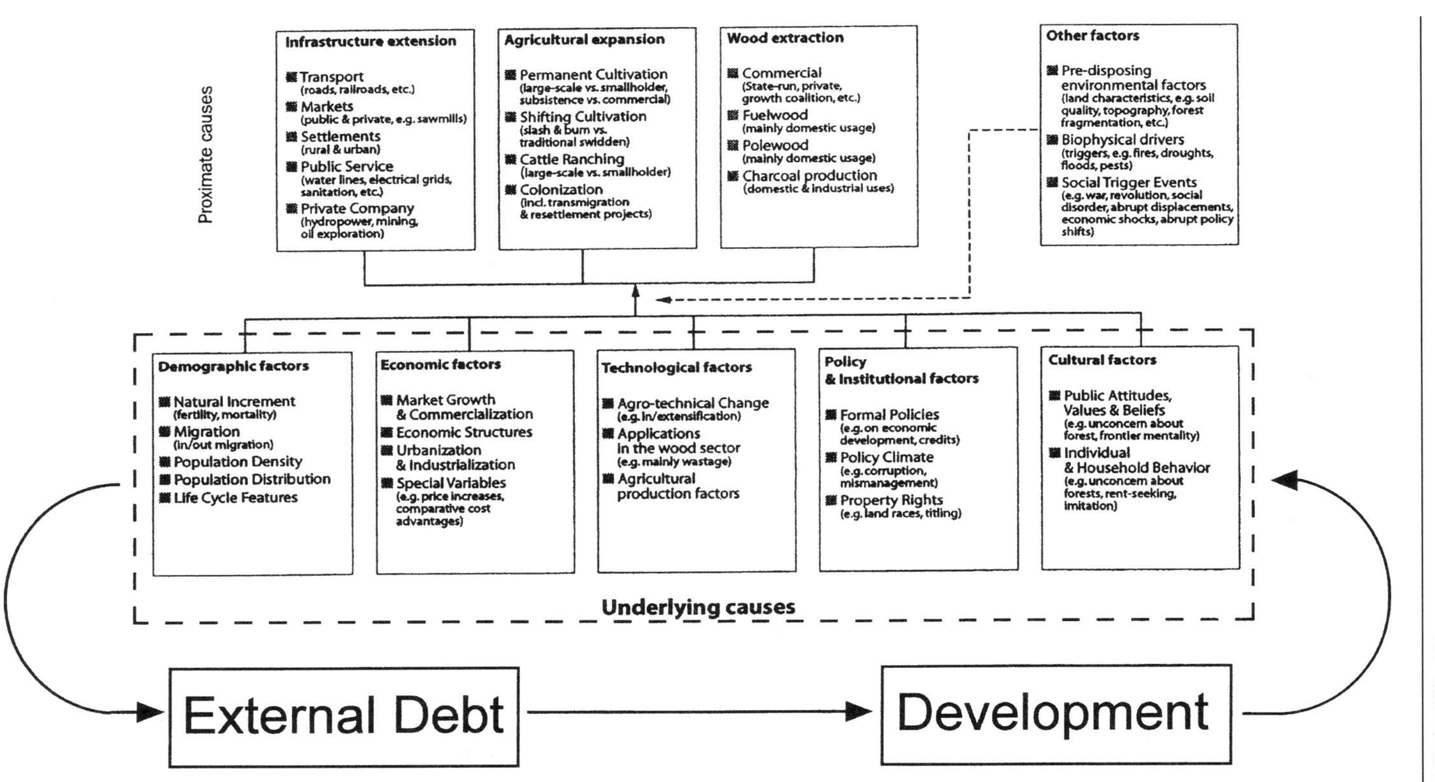

Fig. 4.1. Proximate causes of tropical deforestation (immediate human actions), factors underlying these causes, and forces driving the underlying causes

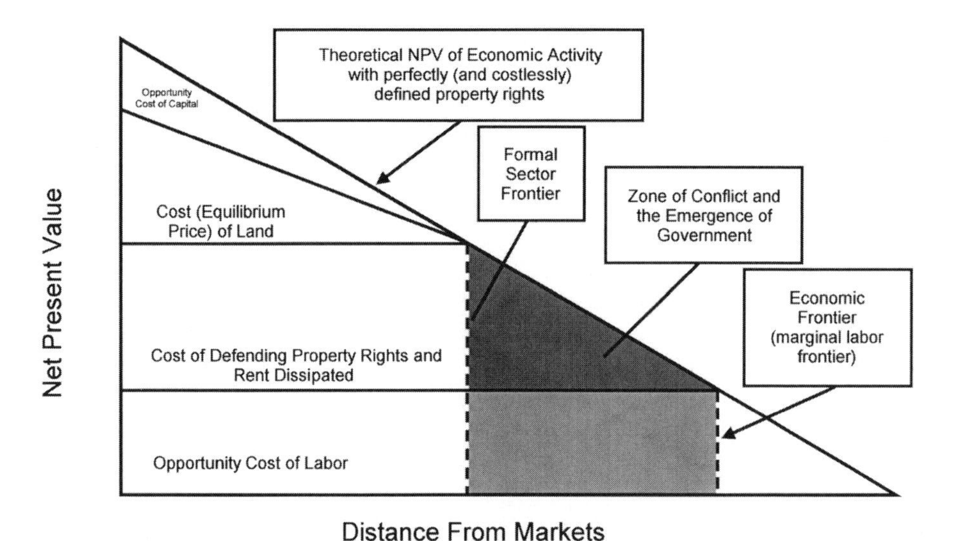

Fig. 4.2. Economic value (*NPV* net present value) of a parcel of land on the frontier as a function of distance from markets where products from that parcel could be sold. (Adapted from Schneider 1995)

Economic frontiers are those areas most remote from markets. Often these are settled by landless peasants who practice shifting cultivation. Such areas are not only distant from markets but also distant from any type of infrastructure such as roads, railroads or small cities. Even if the land is legally owned by a corporation, an individual investor or by the government, the land is essentially "free" to the peasant. The cost of keeping out the squatters is more than the land is worth to the official owners. The only cost to the peasant in such areas is his and his family's labor.

As infrastructure begins to develop in such regions, the logistical "distance" to markets begins to decrease, and we enter the "zone of conflict and the emergence of government". As a road is built across a region, property rights begin to be an issue. If the land officially belongs to someone or some institution, they may try to evict the squatter, either forcibly or through negotiations. Obtaining a legal title to a piece of land through "squatter's rights" requires an investment of time and money on the part of the peasant, and many prefer to sell their "rights", or just to give up and move on to the next frontier. Land is given up or sold not for ecological reasons of decreasing productivity (although that may be a factor), but mainly owing to economic considerations. Often, land is bought up by speculators or by companies engaged in speculation hoping to profit from the increase in value brought about by increasing governmental infrastructure in the area (Schneider 1995).

New capitalist owners of the land often do not begin managing their land right away. Costs of importing fertilizers, herbicides, etc. and of exporting crops are still too high for the landowner to use the land profitably. The landowner assumes, however, that the government will eventually build and improve roads into the region, and establish governmental services such as health care, schools, and market support in local villages. Meanwhile the land lies abandoned.

Usually substantial government services do emerge in a frontier region. As a result, the logistical distance to an existing market decreases, or a new market may emerge. At this point, it becomes worthwhile for the owner to begin investing in infrastructure and supplies that will cause the land to yield a profit. The investor may build a ranching complex or a farm, buy trucks and tractors, hire labor, and import agricultural chemicals. At first, profits are small, but as transportation logistics improve, and the market becomes larger and more economically diversified, profits increase. Because the operation is logistically close to the market, transportation costs fall, and the entrepreneur can successfully compete in the emerging market against producers in other regions. Profitability on lands close to the markets is increased by intensification and by specialization. This is not a phenomenon exclusive to the Amazon. Rather, intensification and specialization have been the essence of development in all regions of the world (Barbosa 2000).

4.2.2.2 Political and Institutional Factors

Geopolitical

Geopolitical concerns are often a reason for opening up a frontier, when national boundaries are ill-defined, or territories are threatened. During the rubber boom of the late 19th century, the westernmost region of the Amazon lowlands was disputed between Brazil and Bolivia. Frustrated by the inability of both countries to reach an agreement on their future, people in the region declared their independence from both countries, creating the independent state of Acre (Barbosa 2000). The newly independent state, composed primarily of Brazilians, managed to expel the Bolivians from the area, thus opening the door for Brazil's de facto control. Despite the fact that Brazil was the imperialist power in this case, the possibility of Bolivia reclaiming the region for itself played on Brazilian insecurities.

In the 1970s, the Superior War College of Brazil was influential in promoting a policy for populating and integrating Amazonia with the rest of Brazil. The view was that development policy should follow the geopolitical needs of the country, that is, the population vacuum in the interior should be filled, as a protection against groups who were said to be "communist guerrillas". The reality was, however, that these were local groups protesting the military dictatorship of Brazil, but that was all the more reason for the military to create a series of agencies responsible for development of the country. For Amazonia, the main agency was the Superintendence for the Development of the Amazon (SUDAM). The responsibility for forest protection came under the Brazilian Institute of Forest Development (IBDF), and indigenous people's affairs came under the National Indigenous Foundation (FUNAI). However, these agencies were compliant when powerful interests demanded the clearing of forests or the removal of indigenous peoples. Things began to change as a result of the re-establishment of democracy in the mid-1980s. The new civil liberties allowed indigenous peoples to organize and form coalitions with environmentalists and grassroots organizations within Brazil and abroad. Central to this change has been the participation of a free Brazilian media eager to cover events taking place in Amazonia, sparked to a large extent by national and international interest over the fate of the forest (Barbosa 2000).

Land Tenure

Land tenure policies that give settlers the right to land only if the forests are cleared stimulate deforestation. In many tropical countries, forest clearing is considered to be an activity that "improves" the land, and lack of improvement indicates that the settler has no interest in the land, or is incapable of using it "productively". A common tactic in the Amazon to ensure tenure is to clear a large area of forest, burn it, put a strand of barbed wire fence around it, and stock it with a few head of cattle. There is no expectation, however, on the part of the settler to make any money from the cattle. Rather the expectation is that the land will be sold at a large profit after the government extends roads and facilities into the region (Schneider 1995). Land speculation, especially by corporate ranches, still occurs in the Amazon (Smouts 2003). In the Malaysian state of Sabah, laws dating from the British colonial period make the state government the holder of all forestry property rights, but permit any native person to obtain title to forest land by clearing and cultivating it. In the Philippines, land claims predicated on forest clearance involved not only small-scale shifting cultivators, but also extensive livestock operations (Repetto 1988). In recent years, as a result of the disappearance of the forest frontier in the Philippines, agricultural intensification has become more important than land clearing (Coxhead et al. 2001).

Tax Policies That Encourage Deforestation

A variety of government policies (investment incentives, credit concessions, tax provisions, agricultural pricing policies, and the nature of lease or sale of forest exploitation rights) create incentives to engage in faster deforestation. Such policies are often instituted when the free market does not accomplish geopolitical goals in frontier regions at a fast enough rate. For example, when the Brazilian government in the early 1970s wanted to open up the eastern Amazon region, officials in charge of regional planning and local financial institutions made considerable effort to convince entrepreneurs to invest in the region. Livestock production was publicized as the most promising investment in part because of national demand and because it seemed to carry little risk. Ranching received the highest priority ranking of projects by government agencies, but among ranchers and corporate groups, livestock was recognized as an only marginally profitable enterprise. However, the enormous gains in land value and the use of incentive monies and reductions or elimination of taxes made clearing of forest for ranching a sound financial option (Hecht 1982). As a result of international pressure to conserve rain forests, Brazil reduced or eliminated many tax benefits deriving from deforestation. However, ranching has not declined in importance because cattle production has become increasingly profitable due to improvements in technology, such as grasses that respond well to fertilization (Cattaneo 2001).

Lack of Enforcement of Environmental Laws

While many developing countries have environmental laws to regulate logging, the laws are often weak and have poor enforcement (see case study below on Indonesia). As a result, there has been a recent dramatic increase in foreign investment in tropical logging by companies to take advantage of this situation (Laurance 2000).

The Transmigration Program in Indonesia (Box 4.1) is an example of how several political and institutional factors interact to cause deforestation.

Box 4.1

The Indonesian transmigration program (World Bank 1988; Muntingh 1997; Katoppo 2000; Kusumaatmadja 2000; Fuller 2003)

Resettlement programs were begun in Indonesia in 1905 when Indonesia was under Dutch rule. Because there were high population densities on some islands and sparse populations on others it was believed the resettlement programs would result in better conditions for both the source and the receiving islands.

Beginning in 1979, the scale of the transmigration program significantly increased, from about 52,000 families during the previous 5 years to 366,000 families during the subsequent 5-year period. Movement was chiefly from the overcrowded islands of Java, Madura, Bali, and Lombok to the largely forested islands of Sumatra, Kalimantan, Sulawesi, and Irian Jaya. An important reason for the transmigration program was the limited land available for farming. In Java, most farming families had less than half a hectare of agricultural land. Farmers had to move onto steep slopes and into forest reserves, where cultivation and erosion caused environmental degradation, siltation of reservoirs, and flooding. Urbanization in Java's major cities was occurring at more that 4% per year, resulting in about 1.6 million new residents each year.

Transmigrants were recruited in rural areas. They were required to be married, to be "of good character", and to have had previous farming experience. Migrants were moved by plane or bus, and on arrival at the receiving island they were given a small house on 0.25 ha of village land and 1 ha of cleared land outside the settled area. In addition, they were supplied with planting materials for minor tree crops such as coffee and small livestock. Public facilities including schools and clinics were provided in the village center. Subsistence supplies were provided for 1 year while the land was tilled and crops established. Settlements were expected to be self-sufficient at the end of 5 years.

In 1985, the World Bank, which helped finance the transmigration program, initiated a review of the resettlement program. Their report noted serious deficiencies. Land clearing was often of poor quality, road construction and maintenance standards were low, and the supervision of contractors uneven. Land for settlement became difficult to find, and land for large-scale settlements in Sumatra was virtually exhausted. The provision of agricultural supporting services, including input supply, extension, and credit, was inadequate, and no progress was made on programs to introduce tree crops to existing settlements. Institutional arrangements for coordination, planning, budgeting, and monitoring and evaluation were weak. The most sensitive issues were social and environmental. Rapid land clearing without adequate planning led to conflicts between transmigrant colonists and local people. Deforestation was also noted as an increasing problem. In January 1986, the government of Indonesia made significant reductions in all development budgets in response to declining oil revenues. In May 1986, the budget was further reduced to 38% of the previous year's figure. It was assumed that most of the transmigration in future years would be from unassisted migrants.

At the time that the transmigration program was expanding, concessions were increasingly granted for logging in the outer islands. Through road building, the transmigration program was instrumental in opening up the outlying islands to exploitation and environmental damage. In 1997, huge forest fires raged across Borneo (Kalimantan). Muntingh (1997) detailed how the fires were caused in part by logging companies acting illegally. Logging continued unchecked, even in designated conservation areas, by concession holders linked to the timber industry and politicians. Substances used in the preservation of logs polluted the rivers and water supply, leading to the death of fish and protected animals, including orang-utans and bears.

Logging and its aftermath severed traditional peoples' ties with their customary land, undermined their sense of identity, eroded their religious and cultural framework, and created great disorientation. Traditional communities became alienated and cultural degradation was manifested in violence. In September 1999, an Update Conference on recent developments in Indonesia was held at the Australian National University that reported on the environmental and social developments during the previous decade. The country had become subjected to a regime of "runaway rent seeking, crony Capitalism (contracts given to old friends, sometimes called 'cronies', and relatives instead of opening bids to the free market), nepotism, and blatant corruption". As public outcry increased, the government responded with increased repression. Environmental rules were disregarded, warnings were ignored, and international protests were unheeded as major companies with close links to highly placed officials began to invest in large-scale agribusiness ventures, such as oil palm plantations. Land clearing was carried out during the worst drought in 50 years. An area of 1.7 million ha of forested land was lost to fire. Other costs due to the massive burning included haze-related losses in the transportation sector, disruption to the distribution system, and long-term health consequences to 20 million people exposed to thick haze for over 4 months in Sumatra and Kalimantan. The increasing population in Indonesia's outer islands continues to threaten forests and their wildlife.

A recent study based on satellite and field-based analysis (Curran et al. 2004) has reported that between 1985 and 2001, Kalimantan's lowland dipterocarp forests that had been designated "protected areas" declined by more than 56% (>29,000 km^2). Deforestation resulted primarily from intensive logging by timber concessions, followed by the clear-cutting of residual stands for oil palm plantations. Threatened nomadic and large vertebrates with extensive lowland ranges are predicted to decline precipitously, especially carnivores, ungulates, and primates (e.g. the Malayan sun bear *Helarctos malayanus*, the bearded pig *Sus barbatus*, and the orangutan *Pongo pygmaeus*). Curran et al. (2004) concluded by citing a World Bank report that indicates that rate and extent of loss of lowland protected forest area in Kalimantan far exceed previous projections. Stemming the flow of illegal wood from Borneo requires international efforts to document a legitimate chain-of-custody from the forest stand to consumers through independent monitoring.

4.2.2.3 Technological

Deforestation increases as the technological means to do so become more available (e.g. chain saws, bulldozers, tractors). At the Jarí plantation in the state of Pará in the Brazilian Amazon where the native forest was replaced with plantations of fast-growing tree species, a more efficient way of getting rid of the native forest was needed. The solution was to fasten a chain to two giant bulldozers, a hundred meters apart. As the bulldozers plowed in parallel lines through the forest, the chain pulled down all the trees in the path (C.F. Jordan, pers. observ., 1984).

The development of agricultural technologies also has contributed to deforestation. Early supplies of palm oil, used to make margarine, cooking fats, and soap came from wild oil palm trees in the rain forests. However, as demand increased, forests were cleared, especially in Southeast Asia, for palm oil plantations. Large monoculture plantations linked to nearby processing plants are needed for efficient production. A typical estate in the Malaysian state of Sabah has five 1,000- to 2,000-ha plantation units plus workers' quarters (Grainger 1993).

4.2.2.4 Cultural

Cultural causes of deforestation include attitudes, values, and beliefs toward publicly owned properties, or toward resources that are perceived as being free. Many tropical forests are publicly owned, that is, they belong to the national government. People living in or near the forest often have a tradition of using the forest for their livelihood, and perceive their use of the forest as an inalienable right. Their ancestors have used the forest for generations, and the knowledge of how to use the forest and the biota that it contains is part of their cultural heritage.

Use of publicly owned resources becomes a problem when the demand for those resources exceeds the capacity of the environment to produce them. Hardin (1968) used the phrase "the tragedy of the commons" to encapsulate what can happen when commonly owned resources are overexploited. "Ruin is the destination toward which all men rush, each pursuing his best interest in a society that believes in the freedom of the commons. Freedom in a commons brings ruin to all" (Hardin 1968, p. 1244). This thesis has been criticized because the access to many commons is not open to all, but limited, either by social pressure or legal restraint, to specific groups (Acheson 1987). However, forests in much of the tropics are, in fact, of open access. Most governments do not have the ability to protect these resources, so, in effect, they are free for the taking. While there are often governmental rules regulating the cutting of forests, they are difficult to enforce, and the common perception is that they will not be enforced.

The exploitation of the forest commons by traditional peoples is a cultural trait. However, their use of "free goods" leads to what could be called a feeding frenzy. Ranchers, loggers, miners, speculators, and shifting cultivators, once they see that others have access to the forest, reason that they also should have this right. The economically rational thing to do is to make a claim to land, and the way to do that is to cut down the trees, because that, according to the law in many tropical countries, is what constitutes "making the land productive". Each man is locked into a system that compels him to cut down more forest. This is hardly a traditional cultural trait, but is rather an example of human response to the global culture of acquisitiveness.

Tolerance of corruption is also a cultural trait. Bribery and corruption are severe problems because forest resources are often controlled by a few powerful individuals or clans that regard logging as an opportunity for personal enrichment. In the Philippines, for example, it is common knowledge that President Marcos' family acquired extraordinary wealth by selling logging concessions to foreign companies (Stone and D'Andrea 2001).

4.2.2.5 Demographic

Increases in population result in increasing pressure to open up forested lands to agriculture. The increases can come from high population growth of settlers already in the region or from migrants from overcrowded cities (Castellanet and Jordan 2002). Sometimes such migrations are subsidized by governments responding to pressure to open up forested areas as a relief valve for overcrowded regions of the country. The opening of the Trans-Amazon highway in Brazil was motivated in part to give drought-stricken farmers of Brazil's northeast access to Amazonian lands. The Indonesian Transmigration Program described in Box 4.1 is also an example. Migrations also occur when small farms in one region of the country are appropriated by corporate agriculture, and the newly landless farmers must be resettled. For example, the migration of landless farmers to Rondónia in the western Amazon was prompted by the development of soybean mega-farms in southern Brazil (Jordan 1995b).

4.2.3 External Debt and Deforestation

Just as the proximate causes of deforestation do not operate in a vacuum, but rather are stimulated by underlying causes, so the underlying causes do not just appear of their own accord. They result from the desire of "lesser developed countries" (LDCs) to become developed, and from the loans extended to the LDCs by developed countries to facilitate this process (Fig. 4.1). Once the LDCs step onto the development/external debt treadmill, there is no getting off. It has been a vicious cycle with development resulting in debt, and the only way to relieve the debt has been through more development. In the 1970s and 1980s, many developing countries overborrowed relative to their ability to repay the debt. In some countries, debt service obligations have been so large that they exceed new loans and private external investment. External debt resulting from importation of international goods can be considered to be the principal cause of deforestation in developing tropical countries (Kahn and McDonald 1995), in as much as the development needed to repay the debt stimulated the underlying causes of deforestation, which in turn drove the proximate causes.

Many different options are available to a country to deal with its debt problem, such as debt rescheduling, debt repudiation, increased borrowing, and restricting imports and increasing exports. One option to repay debt for a country with large areas of tropical forests is to cut down the forest and export the timber. Each 1 million US dollars in external debt was associated with 8.4 ha of deforestation in Asia and 27.2 ha of deforestation in Latin America (Dorman 2003).

Liquidating forests is very attractive to governments, because removal of forests does not show up on a nation's system of national accounts. The forest is not considered to be capital, therefore its removal is not counted as loss of capital (Jordan 1995b). It is as if this resource were free. Because the forest resource is considered a "free good", it is usually used in a non-optimal way, that is, used wastefully. This failure of the national economic system to value resources leads to impoverishment of the resource without the government and the international banks and lending agencies ever becoming aware of the impoverishment.

It is inconsistent that the United Nations' System of National Accounts (Repetto 1992) considers the depletion of oil as a nation's capital depletion, but does not consider removal of forests as any type of depletion. This may be a legacy from the outdated consideration of resources as either "non-renewable" (for example, oil) or "renewable" (living resources). However, it is very unlikely that tropical forests are renewable, because the scale at which they are being cut down eliminates the mechanisms that enable them to be renewed.

4.3 Effects of Deforestation

Chapter 1 presented the values of forests in terms of timber resources and fuelwood. It also referred to non-timber tropical forest products such as pharmaceuticals, fibers such as rattans, and foods such as Brazil nuts. Also discussed were non-market values of tropical forests such as preservation of biodiversity, regulation of climate, carbon sinks, indigenous cultural knowledge, and esthetics. All of these values are affected as a result of deforestation. Chapter 2 showed how deforestation results in the loss of productive potential of the soil, as a result of nutrient loss following destruction of the nutrient-conserving mechanisms of the intact forest. Here, we present case studies of the impact of deforestation on the environment, and on the people that live in and around tropical forests.

4.3.1 Environmental Effects of Deforestation

Chapter 1 discussed in detail the most important environmental values of forests: preservation of biodiversity, regulation of climate, and carbon sinks. We place emphasis on the numbers of species in tropical forests and how deforestation is causing species extinctions. The effects on climate were also discussed at both the regional and global levels; for the regional level, we discussed the influence of large extensions of tropical forest on the hydrologic regime, and effects of deforestation on rainfall. For the global level, we discussed one of the most important roles of forests, the absorption of carbon from the atmosphere, thereby reducing the atmospheric buildup of carbon dioxide.

Other environmental effects of deforestation refer to changes at the landscape level, including transformation of the landscape from forest into scrubland, with resulting soil loss, siltation of rivers, and effects on wildlife. Forest cutting due to mining of resources leads to environmental degradation due to forest destruction and to contamination of waters and air as a result of the mining activities. The development of iron mines in Carajás, Brazil (Box 4.2) is a case in point.

Box 4.2

Iron mines in Carajás, Brazil (de Almeida 1986; Hall 1989; Barbosa 2000; Carl Jordan, pers. observ., 1987–1989)

The 1964 military takeover in Brazil marked a turning point in policy-making for Amazonia. Until then, public initiatives to exploit the region's resources had been piecemeal, narrowly focused, and inconclusive. Possibly because the military authorities were more sensitive to the geopolitical importance of integrating the Amazon Basin into the national economy, a new and more aggressive development strategy began. There were Brazilian nationalist fears over the intentions of suspected predatory foreign interests toward Amazonia. There was also the desire of some planners to open the region as a relief valve for peasants from northeastern Brazil, suffering from periodic droughts and an unequal land tenure system that forced them off the land.

While many projects were included in the new development strategy for Amazonia, including the Trans-Amazon highway, the Greater Carajás Program was the largest comprehensive development scheme ever undertaken in an area of tropical rainforest. It was officially inaugurated in 1980, 13 years after the largest known high-grade iron-ore deposits in the world were discovered in a region inhabited mainly by rubber tappers, Brazil nut collectors, and indigenous tribes. The project was based on export-oriented mineral exploitation and associated industrial activities to generate a trade surplus and help service the country's mounting debt. Loans for the project came, in part, from European and Japanese banks, and were tied to a contract that guaranteed 13 million tons of iron ore to be delivered annually to foreign smelters. To deliver the ore from the mines to the shipping port of São Luis on the coast, 900 km of railroad was built that carries trains having 160 freight wagons several times a day.

Part of the plan was to begin processing of the ore in Brazil itself, and so some 30 pig-iron smelters and industrial plants were planned near the railroad. To supply charcoal for the iron-ore smelters, 1,800 km^2 along the railroad was to be set aside for eucalyptus plantations. Another major part of the project was the construction of a hydroelectric dam on the Tucurui River, to supply power for processing the 2.2 billion tons of bauxite reserves in the region. Although there was a delay of several years, while Japanese and American interests competed for control of the project, eventually an accord was reached to share the operation with a Brazilian company, and the project began operation in 1984.

While mining was the core of the project, agricultural, livestock, and forestry enterprises extended the project out over 900,000 km^2, the size of Britain and France combined. The Carajás Program transformed the social and economic landscape of the region. It attracted into eastern Amazonia thousands of construction workers in search of employment, gold panners in search of riches, small farmers in search of land, ranchers in search of pasture, and speculators in search of quick profit. Local towns have experienced population increases of 400–800%. Competition for land has led to violence. Ranchers and speculators who supposedly bought huge tracts from the government hired gunmen to drive off indigenous peoples and small-scale farmers who had previously claimed the land. The range wars have continued into the beginning of the 21st century. The consolidation of land holdings has resulted in worsened food security. In the towns there is urban poverty, high levels of unemployment, infant mortality, and malnutrition. The 1 million tons of charcoal required to fuel the local smelters require the removal of over 5,000 ha of forest annually. According to the 2004 website of the company that manages the mine, Companhia Vale do Rio Doce (http://www.vale.com.br/), they are now searching for ways to protect the environment and the indigenous communities, and have established a community relations program in education and social welfare. However, the company does not have authority over the Greater Carajás region, that is, the land surrounding the mine, railroad, and hydroelectric facilities. It is not clear to what extent deforestation is still occurring in the Greater Carajás region, but social conflict is still rampant. As of 2002, a trial was still underway for 155 military police who surrounded 1,500 rural workers who were encamped in the Municipality of Eldorado do Carajás near the mines and killed 19 and wounded 69 of them (http://www.labournet.net/world/0106/mst1.html, http:///www.mstbrazil.org/action030102.html). Although the Carajás project may result in short-term economic gains by industry, the ecological and social consequences of this pattern of forest use will be catastrophic. The combined effects of industrial, agro-livestock, and lumbering activities are turning a large part of eastern Amazonia into an unproductive scrubland, leading to soil erosion, compaction, leaching, a greater frequency of flooding, siltation of rivers and dams, pasture degradation, and atmospheric pollution.

4.3.2 Social and Economic Effects of Deforestation

4.3.2.1 Effects on Indigenous Peoples

Many of the development projects that began in the years after World War II gave little thought as to how such projects would affect local populations, other than assuming that they would benefit. While attitudes in development agencies have changed in recent years, the belief still remains that indigenous peoples can adapt quickly, once they are presented with the benefits offered by developed societies. The following case studies show that this assumption is not necessarily true (Box 4.3).

Box 4.3

Development and deforestation in the Philippines (Westoby 1962; Jordan 1995b; Juan Pulhin, cited in Stone and D'Andrea 2001, p. 65; Center for Resources and Environmental Studies 2003).

In 1962, the Food and Agriculture Organization (FAO) of the United Nations issued a report stating that developing countries had neglected their forest assets that could be converted into powerful engines for economic advancement. Harvesting these resources more aggressively, said the report, would result in a symbiotic relationship between industrial nations and less-developed countries (LDCs). The LDCs which had abundant forest resources could mobilize them for development, in as much as they could be assured of markets within more advanced trading partners. These more developed countries would also benefit from the relationship through the steady supply of forest products, particularly timber, to fuel and sustain further economic development. This was the prevailing attitude among international development authorities in the 1950s and 1960s. As a result, industrial forestry was undertaken in countries such as the Philippines. Contracts and development plans were administered from the top down, an approach that conformed to the accepted economic developments standards of the time.

Stone and D'Andrea (2001) chronicled the impact of such development on the upland forests of Mindanao, and on the indigenous Lumads who occupied these forests. For centuries, the Lumads lived on the upper slopes of Mindanao in Bukidnon province, in relative isolation. Following World War II, the population of the Philippines exploded, and after the best agricultural soils in the lowlands were occupied, migrants from other islands began moving into Bukidnon. The Lumads fled to more remote areas, because they had no land rights. However, the law of the land allocated these "undocumented" lands to the public domain. The Lumads were considered squatters, and had to move on when a number of foreign logging companies began operating in the province. By the mid 1970s, these companies, and the national ones that eventually replaced them, had stripped Bukidnon of its timber. The companies then moved on to new territories. The number of forest licenses issued to timber companies and the number of hectares felled each year reached a record between 1972 and 1984. By 1990, only 20% of the country was forested, and most of this was second-growth forest with less commercial value or biological importance than the old-growth forest.

Some of the immigrants to Bukidnon were able to establish farms in the cut-over forests, but periodic fires converted much of the land to alang alang (*Imperata cylindrica*), an aggressive grass that competes with crops and that is very difficult to control. The hoped-for benefits of forest-based development had not materialized. Fortunes were made as forests were devastated. However, nearly all the benefits were funneled to those in political power, with practically no benefits to the local farmers (Stone and D'Andrea 2001).

The Tala-andig are another indigenous group that lived in the mountains around the central plateau of Mindanao. Like the Lumads, they were forced to higher elevations as logging companies cleared the valley, once covered by moist forest. Now they live on lands officially claimed as "ancestral heritage", but the forests still are endangered. Soils cleared of forest at high elevations are ideal for growing potatoes, much in demand in Manila fast-food restaurants. However, they can be grown for only 1 year before suffering from wilt, so the migrant farmers continually move upslope, seeking fresh ground. Forests are opened up through "accidental" fires in alang alang, which quickly spread upslope.

There are many more cases of where governments of tropical countries have granted logging concessions to national or international companies with little thought of how deforestation would affect the indigenous peoples that live in the regions. In many cases, such as that of the BaNgombe and BaKoule of the Congo (Box 4.4), loggers removed only a small proportion of the trees. Nevertheless, the presence of the logging company had a large impact on these people, both economically and socially. In other cases, such as that of the Penan in Borneo (Box 4.5), environmental disasters such as fires resulting from careless logging added to the impact on indigenous peoples.

Box 4.4

The BaNgombe and BaKoule of the Congo

Wilkie (1996) discussed some short- and long-term consequences of commercial selective logging by the Société Forestière Algéro-Congolaise (SFAC) on indigenous BaNgombe foragers and BaKouele farmers of the northern forests of the Republic of Congo. SFAC is a semipublic company formed in 1983 within the framework of a 20-year cooperative agreement between the Democratic and Popular Republic of Algeria and the Republic of the Congo. In 1985, SFAC began selective logging in a concession of 855,000 ha in the Sangha region.

Before the colonial period (1900–1960), the BaNgombe lived a seminomadic existence. They had a long-term exchange relationship with BaKoule farmers who were settled in villages alongside perennial rivers and streams and practiced rotational slash-and-burn agriculture. The BaNgombe traded farm labor and forest products such as meat and honey for cultivated crops and commodities such as salt, clothing, and tobacco. They traveled in the forest for extended periods of time and hunted forest antelope and primates with crossbow, nets, and traps.

The arrival of SFAC had a profound effect on the local economy. Although the company used heavy machinery to build roads and transport cut logs, the company still had to hire a relatively large number of workers on a daily or monthly basis. The tribal compositions of work teams had clear differences. Inventory and exploitation teams were primarily BaKouele and BaNgombe, because of their intimate knowledge of the area and the tree species. The drivers and mechanics were usually from regions outside the Sangha. Despite their low wages, BaKouele and BaNgombe families with SFAC employees were more likely to have tin-roofed huts, new aluminum cooking pots, eating utensils, flashlights, and new clothes and shoes than those families with no SFAC workers. The logging operation improved health-care services, primary education, and housing conditions for a small portion of the BaKouele who lived close to headquarters. The BaNgombe, however, did not benefit, owing to implicit prejudice in the allocation of worker housing. As a result of the entrance into the cash economy, artisan and hunting skills were lost. SFAC employees used wages to finance commercial wild game hunting by buying shotguns and cartridges.

Although SFAC logging spurred the local economy and enhanced the material quality of life for BaKouele and BaNgombe employees, its effect on social services was very local. In addition, the economic income was unlikely to last, as the concession was only for 20 years, at which time most of the valuable timber in the area will have been taken. In the conclusion to his paper, Wilkie (1996) asked: "How will five to ten years or more of high income affect the needs, aspirations, and social behavior of a given local population of BaNgombe and BaKouele? Will today's employees be able to return to a more basic lifestyle once SFAC moves out of their region? Will they be able to re-attain traditional sharing patterns? Will young people have neglected to learn traditional techniques that once again become more important to daily subsistence? Will the state be able to assume the role of education and health care provider once SFAC moves out of an area? Will faunal populations be able to recover from such intensive market hunting and thus continue to provide the local population with a necessary source of protein?"

For the BaNgombe and BaKouele, the long-term costs may outweigh the short-term benefits.

Box 4.5

The Penan of Borneo

During the 1980s and early 1990s, Dr. J.P. Brosius conducted research among the Penan, an indigenous group that has inhabited the interior of the Malaysian state of Sarawak. One of the objectives of his work was to characterize how the government was able to overcome the resistance, not only of the Penan themselves, but also of international environmental organizations, to deforestation of the Penan homeland (Brosius 1995).

The natural landscape cover of Sarawak on the island of Borneo is moist forest. For the Penan, the landscape has been more than simply a reservoir of detailed ecological knowledge or a setting in which they satisfied their nutritional needs. A strong coherence has existed for them between the physical landscape, history, genealogy, and the identities of individuals and communities. Rivers were the paradigm around which spatial, historical, and genealogical information was organized.

The Penan were migratory, and the factor that more than any other determined their movement was the availability of the sago palm, *Eugeissona utilis*. Sago has been the principal source of carbohydrate for the Penan. Trunk sections were split, and the pith was pounded, rendering it soft and pliable. The pith was then placed in baskets and trampled by women, while pouring water through. The starch was separated from the pith by the trampling, and it washed through into the settling mat below.

Sago can be harvested sustainably. It reproduces vegetatively if the roots are left intact. When the sago in one area was depleted, the people moved to another area, leaving the previous stand to recover. The Penan ethics of resource use was one of explicit stewardship. Leaves of other palms were used for weaving. Women were the exclusive weavers for the Penan. Their mats and baskets were in great demand by other groups who in turn traded with Chinese merchants.

Penan hunters used blowpipes for small game such as monkeys, squirrels, and barking deer. For their blow darts, they used poison extracted from a local tree. For larger game, dogs were used to chase and corner the prey until it could be speared by a hunter. The favored game was bearded pig, not only because of its large quantity of meat, but also because of its substantial deposits of fat that could be rendered and stored for later use.

The first signs that the sustainable coexistence of the Penan and forest was ending were the survey markers for logging tracts. By 1992, a bridge and logging road had driven the game out of the Seping River Valley in the center of the homelands, and clogged the river with mud. Clearing of the forests changed the perspective of the rivers to the Penan. The cultural symbols were no longer recognizable. Hunting grounds were destroyed, as well as ancestral burial grounds. Trees that had little market value were cut and discarded, even though they had great value to the Penan for making tools and blowpipes.

The Penan responded by erecting symbolic barricades in the forest. Their protests attracted the attention of international organizations such as the Rainforest Action Network, who organized a series of blockades that galvanized global concern. The Malaysian government responded in various ways. One was to trivialize the issue. They acted toward the complaints as one would act toward the complaints of wayward children. "Authority knows best, and what it does is for the good of those disciplined". This attitude was especially apparent in the bemused and contemptuous attitude of loggers toward the Penan when they encountered them in the forest. The government did begin programs to help the Penan, such as giving them sheets of plywood for their shelters. For the most part, however, the government rejected emotional scenes by claiming that the scenes were instigated by "imperialists". The Sarawak chief minister summed up the government's attitude as follows: "How can we have an equal society when you allow a small group of people to behave like animals in the jungle... I owe it to the Penans to get them gradually into the mainstream so that they can be like any other Sarawakian".

In 1992, the Earth Summit meeting in Rio de Janeiro called for the world's greater attention to complaints of indigenous groups such as the Penan. The Manila-based Asian Development Bank (ADB), which since 1977 had been making forestry loans, accordingly began to develop new policies with more environmental safeguards. Its 1995 forest policy paper stressed "the need to balance the three imperatives of production, protection, and participation". As a result of this policy, there began an effort to include the participation of local peoples in plans for the forest. However, the paper did not rule out bank support for plantations and production forestry.

In 1997, smoke from the fires in Borneo clouded skies, closed airports and entire communities, and provoked complaints from as far away as Singapore and Kuala Lumpur. London's *Sunday Times* classified the disaster as "wholly man-made" – because many of the hundreds of fires were started deliberately as a cheap way to clear land by companies with corrupt connections to government officials. Shortly thereafter, the Asian Development Bank's principal environmental officer admitted in an interview that "we are still evolving our strategies for participation" (Stone and D'Andrea 2001).

Costs and Benefits

There really are no real benefits to indigenous peoples as a result of the deforestation of the land in which they live. Often, they are paid small sums of cash by the government or by the corporations that take their land, but, soon afterward, both the forests and the money are gone. Sometimes the tribes are moved to another area, but almost always their situation will deteriorate, either because they are not familiar with the environmental situation or because they come into conflict with other groups that are already there. In some cases, the people migrate to cities where they live in urban slums, because they do not have the skills to compete in the modern economy, or because the cities are not prepared to absorb the extra manpower or to provide adequate living conditions for the migrants.

In recent years, some indigenous groups have learned to use the tactics of the antiglobalization movement to agitate against national policies that they see as destructive to their environment and culture. For example, in the south of Chile, the Mapuche Indians have become major political players fighting timber companies who want to exploit the ancient forests that constitute their tribal homelands. In Bolivia, radical Indian leaders seized upon a dispute over tribal justice to mobilize thousands of protestors. They blocked roads and laid siege to La Paz. As a result, the president, Sanchez de Losada, fled into exile in the USA (de Cordoba 2004) and on 17 October 2003, Carlos Mesa assumed the presidency of Bolivia.

4.3.2.2 Effects on Traditional Rural Peoples

The effect of deforestation upon traditional rural peoples is much the same as that upon indigenous peoples. Both have lived in the forest for generations, have come to depend upon the forest for their livelihood, and have practiced management that while not always ecologically sound, nevertheless did not destroy the regenerative and productive capacity of the forest. Deforestation in India (Box 4.6) serves to illustrate this point.

Box 4.6

Deforestation in India

Alcorn and Molnar (1996) described the conflict arising from deforestation in India. As in most developing countries, deforestation affects two interest groups: commercial interests and subsistence interests. The commercial interest group has used forests to generate capital, as if nature were just another asset to be converted into some other capital asset without penalty. On the other hand, members of the subsistence interest group view forests as the base of their support. Destruction of the forests means the end of their benefits.

Through its rules, policies, and price supports designed to promote industrialization, as well as through budget allocation and economic analyses, in India the state has generally supported the commercial interests allied with the political elites. Those dependent on nature for subsistence have exercised little political power. Although Indian communities have long fought to retain or regain rights to make decisions about forest management, the state has usurped their rights in the name of modern management and conservation.

The subsistence base of three major sociopolitical groups in India has been particularly affected by this progressive loss of rights and alienation from forest management: pastoralists, tribes, and sedentary farmers. Pastoralists are largely dependent on open woodlands for fodder to supplement pasture. Tribes in India have been concentrated in the hill forests, especially at the northeastern border with Myanmar and China. Tribal cultures had the most rules regarding forest management, ranging from replanting to the maintenance of sacred groves. The third group is sedentary non-tribal villagers who depend on forests for fodder for their livestock, for cooking fuel, for timber, and for non-timber forest products. This group has had the most success in resisting usurpation of their forests, in part because of the Gandhian tradition of peaceful resistance and a religious tradition that values peace for all living organisms. The "Chipko" movement, in which women hugged trees to keep them from being felled for commercial use, began in 1972 and in 1981 achieved a ban on commercial deforestation in an area of 40,000 km^2.

Nevertheless, deforestation continued to be a major problem in India in the 1990s. In the industrial sector, shortages of raw materials and obsolete equipment caused forest-based processing enterprises to operate at a fraction of their capacity. The demand for pulp, paper, and manufactured wood products spiraled as urban and middle-class incomes rose and consumer demands for wood and paper products increased.

Beginning in the 1970s and 1980s, there have been efforts to combat deforestation. Almost all the Indian states have had extensive afforestation programs to meet the rising demand for forest products and to help check deforestation. Social forestry programs have been implemented following Gujarat Forest Department's experiments with community woodlots in the 1970s. In 1976, the National Commission in Agriculture recommended a national social forestry effort, and government, bilateral donors, and multilateral development banks funded it. They were originally directed at the fuelwood crisis, but have evolved to supply all types of forest products. Evaluations of the program have been varied, depending on the criteria used. While industrial supply of wood or pulp has in some cases met the needs of industry, social equity issues have fared less well. In many cases, poor people were hurt when common property resources were closed to them in order to create plantations whose products have mainly helped those already rich. On the other hand, social forestry programs have frozen the common property status of land, and thereby prevented further privatization.

Hundreds of grassroots groups have arisen in India, and their concern is with conserving the environment for the benefit of local communities. These groups are not concerned with environmental protection per se, but with the proper use of the environment and who should benefit from it. For example, the vision of the "Centre for Science and the Environment" [a major non-governmental organization (NGO) based in New Delhi] calls for each rural settlement in India to have a clearly and legally defined environment to protect and improve.

Over the past few years, there has been a marked change in the Indian economy, as a result of the globalization of some service industries. High-tech firms such as IBM are now setting part of their operations, such as software programing, in India as well as in other developing countries where wages and others costs are much cheaper than in industrialized countries. Medical centers in the USA are relying on medical doctors in India to process data. Many less skilled Indians are manning call centers as service representatives for US corporations (Irwin 2004). Many young people are being trained in telephone etiquette and to speak with regional accents, depending upon where the incoming call originates. These young people are breaking away from tradition, and are increasingly involved in the global economy. The transformation in India has been remarkable. "India has shifted away from socialism and dived headfirst into global trade, the information revolution and turning itself into the world's service center" (Friedman 2004). Whether the new service economy will replace the older extraction economy on a scale to reduce the problem of deforestation and environmental degradation remains to be seen.

4.3.2.3 Effects on Recently Arrived Rural Peoples

Tropical forests and the soils that sustain them represent natural capital, that is, they are a resource that required no human effort to establish. When natural capital becomes available to anyone for the taking, there naturally will be a rush to establish a claim on this open-access commons. The arrival of farmers in the Amazon and the consequent agricultural development in this frontier region (Box 4.7) illustrate the phenomenon.

Box 4.7

Agricultural development by migrant farmers in the Amazon

Before the 1970s, the land around the river village of Altamira on the Xingú River (a tributary of the Amazon) was completely forested, and was occupied only by indigenous tribes and by sparse caboclo (descendants of indigenous and white people) populations along the main river who lived by fishing and rubber extraction. Altamira was the center of public agricultural colonization, opened by the military government in 1972 through public subsidies and government-planned centralized programs. The population grew rapidly during the 1970s, owing to government incentives and propaganda (the public slogan was "Amazonia: a land without men for men without land"). However, the number of farmers who actually settled was much lower, and the costs per farmer much higher than the very optimistic initial forecasts. A large proportion of the farmers, disillusioned with the lack of infrastructure and the low fertility of the soils, abandoned their land after a few years, but were replaced by newcomers, mostly landless migrants from northeastern Brazil. After a few years of poor results, government support was reduced (Moran 1981, 1996). However, the flux of migrants into the region continued spontaneously during the 1980s and early 1990s and later decreased gradually.

In the 1990s, LAET (Laboratório Agro-Ecológico de Transamazônica, or the Agroecological Laboratory of the Amazon region, an NGO funded in part by the European Union) began a project in the region around Altamira to encourage farmers to improve their methods of natural resource management, thereby reducing deforestation (Castellanet and Jordan 2002). As part of their work, LAET members carried out a survey in the Altamira region to characterize the farmers of the region, including their economic status.

In contrast to the impact of deforestation on traditional farmers in India, it could be said that deforestation in the Altamira region actually helped farmers, in that it opened up new lands to be exploited. The critical difference from India is that the Brazilian farmers were not from the region deforested, but rather from other parts of the country. The LAET survey (Table 4.5) suggested seven different categories to characterize farmers:

1. "Just arrived" farmers were those who lived on the land for 4 years or less. The long distance from all-weather roads made it very difficult to market products, and, as a result, profitability was low.
2. Pepper producers did not do especially well, in part because of problems with wilt, a fungal disease caused by *Fusarium*.
3. Of all farmers, cocoa producers were doing best at the time of the survey, but cocoa prices are cyclic, and income was not continually maintained.

4. Some farmers actually were losing some of their original investment ("losing capital").
5. Farmers who diversified (cacao + cattle) were able to maintain themselves relatively well.
6. Cattle producers who had enough capital to improve their pastures or to continually acquire new land for new pastures maintained themselves.
7. Small-scale ranchers (glebistas) who were forced to maintain their herds on pastures that were rapidly losing fertility did poorly.

Despite the government's promise that new lands would provide wonderful opportunities for farmers who would settle the Amazon, the new lands actually benefited the farmers relatively little from an economic standpoint.

After 5 years, LAET researchers concluded that the main problem of resource conservation was not soil fertility, but the extension of poorly managed pastures. As a result, they suggested that a better strategy for the region would be to encourage more intensive management of smaller-sized holdings. A comparison of social and economic factors for farmers with land holdings of different sizes suggested that 25–35 ha should be sufficient to maintain or even increase the level of agricultural production, based on the following cycle: 2 ha of annual crops for 1 year, intercropped with leguminous cover crops, followed by 5 years of pasture, and 5–10 years of fallow before a new slash-and-burn cycle.

Table 4.5. Characteristics of farmers interviewed in the 1994 Altamira survey (Castellanet and Jordan 2002)

Type of farmer	No. of years on the land (ha)	Distance from the road (km)	Income (US$/year)	Percentage of income[a]				Pasture area
				Cacao	Pepper	Cattle	Rice	
Just arrived	4	32	2,015	0	13	1	20	7
Pepper producer	14	15	3,138	7	25	10	8	23
Cacao producer	15	19	7,990	59	8	5	6	37
Losing capital	13	18	1,219	0	4	3	23	5
Cacao + cattle	15	6	6,962	31	13	28	7	38
Cattle producer	16	6	7,562	8	6	40	4	68
Rancher	11	30	1,489	1	0	60	2	197

[a] Other sources of income include small animals, extractivism, commerce, small business, and labor sale (wages)

The Local Elite in Altamira

The overall objective of the LAET (Laboratório Agro-Ecológico de Transamazônica, or the Agroecological Laboratory of the Amazon region) project was to improve management of natural resources, with a specific goal of slowing the rate of deforestation in the region along the Amazon highway near Altamira, a center of colonization. The initial focus group was made up of farmers, since they were suspected of having the greatest impact on natural resource management. However, once researchers began to work with those in forestry and the wood industry, they realized that sawmill owners and large ranchers played a greater role in deforestation, especially when their interests coincided with those of farmers. For example, sawmill owners opened roads or rehabilitated them and encouraged further occupation by landless farmers (posseiros) to cover up illegal logging and to provide cheap manpower and logistical support for loggers. For farmers who were already established, sawmill owners arranged for repair of damaged feeder roads and for provision of free rides to the city.

Big ranchers also had common interests with small farmers. Ranchers offered the farmers opportunities for day labor, assistance in transport, and the renting or sharing of cattle. Some small farmers had a strategy of converting their land from forest to pasture and then selling it at a good profit to ranchers. Ranchers who wanted to expand their pastures quickly without having to depend on contracted manpower often depended on buying land from small farmers.

Merchants and service people in small towns also had an interest in expansion of agriculture, since farmers and ranchers were their main customers and providers of commodities. The business community also favored farmers, because their contribution to the population of the region was important in obtaining support from the state and federal government. Such support was based on the size of the county's population.

Because of the financial interest of these local groups in economic expansion, it was difficult for LAET to stimulate interest in concepts such as zoning that would reduce deforestation. However, there was a difference in attitude between business leaders in villages on the upland frontier, where regulations concerning natural resource management were strongly opposed, and long-established river towns where people were more aware of problems of deforestation, and therefore were more sympathetic towards regulations that would preserve forests (Castellanet and Jordan 2002).

4.3.3 Benefits and Costs of Deforestation at the International and National Levels

4.3.3.1 International

Deforestation and development result in benefits for countries and companies that import tropical timbers. These benefits include: increased business for international timber companies; lower prices for wood; employment for value-added industries such as furniture manufacturers; and overall increase in trade that could help industries that export products to tropical countries. Costs might include the widening of the gap between the have- and have-not countries, if prices paid for tropical timbers are lower than the replacement costs. Pricing timber below replacement costs decreases even further the ability of tropical countries to catch up economically with developed countries. Such economic disparities can result in international tension.

4.3.3.2 National

Deforestation to finance development can produce short-term benefits for the developing country. The benefits can include: easing of debt crisis; obtaining currency for foreign exchange; securing national boundaries; attracting international investment; easing of social problems in some overcrowded areas (due to transmigration programs); and increasing political control in frontier areas (benefit for governing party, not for opposition parties).

However, if the prices received for exported timber are below replacement costs, the solution to debt crisis will only be temporary. Loss of natural capital precludes long-term solution to economic ills; delays long-term solutions to economic problems; increases social problems on the frontier; increases environmental problems; and increases obligations to transnational corporations that may not be in the best national interest.

4.4 Conclusion

Although the rate of deforestation in the tropics is debated, the overall trend is clear. The world is experiencing a significant loss of its tropical forests. The data in Table 4.1 (total forested area divided by area deforested annually) suggest that if the rate does not slow, tropical forests will have disappeared within a time frame of a century. The law of supply and demand predicts that as tropical timber becomes scarcer, its use will become more efficient, and, as a result, the rate of forest depletion may slow. The problem is that, by the time this occurs, the ability of tropical forests to reproduce may be seriously impaired.

As tropical forests disappear, the productive potential of the land is affected, as are local and national economies. For some groups, the adjustments will improve their economic income, while for others their situation will deteriorate. For everyone, there will be a loss of an international treasure of immeasurable value.

Management of Tropical Forests 5

5.1 Introduction

People have managed tropical forests for extracting timber and other resources for millennia, but on a scale small enough not to damage the ability of the forest to recover its original structure and function. Today, many tropical forests of the world are being managed unsustainably, generally due to the intensity of the timber harvest and the lack of adequate techniques to preserve sustainability, that is, to preserve ecosystem structure and function and to ensure the ability of the forest to regenerate populations of desirable tree species. However, many management systems have been designed to avoid damage to the forest structure and to maintain the forest in production. In this chapter we examine management practices for long-term sustainability of natural forests, the way the practices have evolved, and current trends in forest management.

Systems for sustainable management of tropical forests should take into consideration features that are essential for the maintenance of the natural forests' structure and function:

- the maintenance of ecosystem biodiversity, including the mutualisms that are essential for forest reproduction;
- the maintenance of viable populations of wildlife;
- the maintenance of the nutrient retention and recycling mechanisms of the forest;
- the maintenance of soil organic matter.

5.2 Natural Forest Management

Natural forest management has been defined as "controlled and regulated harvesting, combined with silvicultural and protective measures, to sustain or increase the commercial value of future stands, all relying on natural regeneration of native species" (Schmidt 1991). Compared to plantations and agroforestry, natural forest management systems are less intensive, with relatively lower short-term yields, but also requiring lower capital inputs. There is a long history of management of natural forests by indigenous peoples. For example, recent archaeological research in the Upper Xingú region of Brazil has revealed that a complex settlement pattern existed in the Amazon long before the arrival of European colonizers (years 1200–1600 a.d.). These settlements apparently greatly transformed the local landscape and sustained human populations without destroying biodiversity (Heckenberger et al. 2003; Stokstad 2003). Another recent study has shown a history of human occupation in other regions that otherwise were considered to be pristine, such as the Solomon Islands (Bayliss-Smith et al. 2003).

It appears that several cultures have been able to use the forest in a sustainable manner to their benefit. Even today, large areas of tropical forests are used by indigenous peoples. Such forests can be considered "human-dominated ecosystems", since many forests are actually used for gardening, agroforestry, hunting and gathering (Noble and Dirzo 1997). The key to using tropical forests to sustain human populations in indigenous societies has been to conserve the basic forest structure and function. Natural forest management of today is based on this same principle.

Modern natural forest management is often not sustainable, and has been criticized on a number of grounds (Buschbacher 1990; Bruenig 1996; Reid and Rice 1997; Pereira et al. 2002; Frederiksen and Putz 2003):

- It is based on a few valuable species including *Swietenia* spp. (mahogany) in Latin America, *Khaya* in Africa, and *Shorea* in Asia. Selective logging can reduce or extinguish local populations of these species.
- Other species are not used, because existing markets are often not sufficient to warrant their extraction.
- Selective logging practices cause a certain degree of damage to the forest, especially when felling and transportation of logs is carried out on steep slopes. Damage to the forest is highly dependent on the intensity of logging and therefore sustained forest management techniques necessarily restrict the amount of timber harvested.
- Due to the damage caused to the forest by selective logging, there is often not enough natural regeneration.
- Selective logging opens up areas that local people can use for shifting agriculture.
- Logging roads increase access to wildlife hunting.
- Opening of the canopy by logging increases risk of damage by fire.

In the following sections we explain how some of these negative effects can be avoided or counteracted, using knowledge of the basic ecology of natural forests to design sustainable management systems.

5.2.1
Sustainable Forest Management

According to the United Nations Environment Program (UNEP), sustainable development "meets the needs of the present without compromising the ability of future generations to meet their own needs and does not imply in any way encroachment upon national sovereignty". It also implies "the maintenance, rational use, and enhancement of the natural resource base that underpins ecological resilience and economic growth". Sustainable development also implies incorporation of environmental concerns and considerations into developing planning and policies (UNEP Statement on Sustainable Development, Governing Council 1989, in Higman et al. 1999). Sustainable development is economically viable, environmentally benign, socially beneficial, and balances present and future needs (Higman et al 1999).

Sustainable forest management (SFM) is management of natural forests in such a way as to minimize the problems associated with timber extraction. SFM has been described as forestry's contribution to sustainable development. It aims at maintaining the productivity of the forest for timber and other human needs through preservation of soil fertility and hydrological stability. In addition, sustainable forest management maintains levels of biodiversity that occur naturally in the forest (Dawkins and Philip 1998). A more complete definition of SFM is given by Bruenig (1996):

> ...management should aim at forest structures which keep the rainforest ecosystems as robust, elastic, versatile, adaptable, resistant, resilient and tolerant as possible; canopy openings should be kept within the limits of natural gap formation; stand and soil damage should be minimized; felling cycles must be sufficiently long and tree marking so designed that a selection forestry canopy structure and a self regulating stand table are maintained without, or with very little, silvicultural manipulation; the basic principle is to mimic nature as closely as possible to make profitable use of the natural ecosystem dynamics and adaptability, and reduce costs and risks....

Many SFM systems have been designed to follow these guidelines as closely as possible.

In the following section, we describe the principal systems of tropical forest management that have been or are still being used in different regions of the world. Then we focus on some recent initiatives that seek to standardize management techniques with the overall goal of increasing sustainability through the establishment of criteria and indicators of sustainable forest management, along with the development of "better management techniques" for forest management, such as reduced impact logging (RIL).

5.2.2
Systems Used in Management of Natural Forests in Tropical Regions

Some of the systems used to manage tropical forests worldwide are summarized in Table 5.1. Some of these systems are still practiced, while other systems have been discontinued or modified.

5.2.2.1
Natural Regeneration Systems

The success of management systems based on natural regeneration depends on the number of individuals of desirable species left after harvest. The species managed should be abundant, have a wide diameter distribution, and good quality timber. Diameter increments of managed species can be increased by selectively favoring desirable species through refinement and liberation treatments to reduce competition in the stand. Refinement and liberation are used in most systems of natural forest management. Refinement stimulates growth by eliminating the overstory of undesired species and individuals. It consists of the eradication of unwanted vegetation (weeds and defectives) to promote complete utilization of the site by high quality trees of the preferred species. Liberation is the freeing of desirable species from competitors by removing vines, lianas, and other plants that impede growth.

If wildlife conservation is a goal of natural forest management, care must be taken not to eliminate species that are a food source for wildlife. Also, defective trees are often habitat for certain birds like woodpeckers. Given the complex set of mutualisms that exist in tropical forests (described in Chap. 2), it would seem that too much refinement and liberation could "sterilize" the tropical forest by removing important food sources and habitat for

Table 5.1. Systems used in management of tropical forests worldwide

Type of method	Name of management system	Nomenclature	Country of origin or practice
Natural regeneration systems			
Monocyclic methods	Malayan Uniform System	MUS	Malaysia
Polycyclic methods	Celos Silvicultural System	CSS	Surinam
	Selective Management System	SMS	Throughout the world
Clearing systems			
Shelterbelt system	Tropical Shelterwood System	TSS	Nigeria, Ghana
Strip cutting system	Palcazu method		Peru

animals. Most natural regeneration systems lack consideration for faunal biodiversity (Bennett 2000). More recent approaches to SFM which include consideration of faunal biodiversity are discussed later in this chapter. Also, care must be taken not to destroy non-timber forest products (NTFPs) that are important for the subsistence of local populations.

Monocyclic Methods

The Malayan Uniform System (MUS) is one of the oldest and most widely known management system of natural forests in Southeast Asia. It was designed for forests that are relatively uniform and rich in commercial species of the *Shorea* genus in the Dipterocarpaceae family. The genesis and nature of the system have been described by Wyatt-Smith and Panton (1963). All trees of commercial size are harvested in a single operation, followed by poison-girdling of all unwanted stems (non-commercial species, or commercial species with defective stems). Three to five years after the initial harvest, diagnostic sampling is conducted to determine the status of regeneration and to prescribe treatments that will ensure good regeneration. Silvicultural treatments to promote regrowth include climber cutting (liberation) and poison girdling of unwanted trees competing with the more desirable species. These practices are applied at 10, 20, 40, and 60 years and the area is harvested again after 70 years.

In the MUS, regeneration was generally good on lowlands with fertile soils and the system was commonly used until the mid-1960s (Dawkins and Philip 1998). The low cost of mechanical extraction and making and maintaining roads and the lack of demand of other species created an ideal situation for the MUS in the lowland areas. However, competition with other land uses such as rubber and oil palm led to a displacement of this management system to more hilly terrain where conditions were not so favorable for natural regeneration (Buschbacher 1990). Another factor that contributed to the decline of the MUS was the increasing use of wood preservatives which permits the utilization of more species.

By the late 1980s, a selective management system (SMS), similar to one being used in the Philippines, was suggested as a more flexible and appropriate approach to suit the changing conditions of forest management in Malaysia. The new SMS required a pre-felling inventory where permanent sample plots are defined and used to determine growth rates by diameter classes and species groups, mortality, regeneration, and felling damage. The felling regime is then designed based on minimum diameter limits (MDL) for cutting that are defined for each species, so as to conserve the resource, ensure sustainability, reduce damage to the remaining forest, and optimize utilization. Yields of 30–40 m^3/ha were expected in 25- to 30-year cutting cycles, thus transforming the system into a polycyclic method. The system has yet to be tested over time and success greatly depends on the efficacy of the control of logging (Dawkins and Philip 1998).

Polycyclic Methods

The Celos Silvicultural System (CSS), developed by the Wageningen Agricultural University of the Netherlands, has been proposed as a technically feasible balance of economic and ecological aspects of timber production in the seasonal evergreen forest of Surinam (De Graaf and Poels 1990). The CSS is a cost-effective way of growing more marketable wood in previously exploited neotropical rain forest, based on relatively short felling cycles of 20–30 years (De Graaf 2000). It can be used in natural or lightly used forest. The Celos system is designed for areas that are large enough to supply an economically viable timber processing unit. When the area under forest is abundant, an extensive system with low output per hectare is adequate.

The Celos system is based on the use of silvicultural operations in several cycles of interventions. For example, a harvest cycle may consist of an initial extraction of 10 m^3 with subsequent interventions after 8 and 16 years, with a target of a total of 20 m^3 of lumber/ha and a felling cycle of 20 years. Refinement and liberation are used as needed to stimulate the growth of desired individuals that are left as residuals.

The method is good for areas with relatively large tracts of forest thus permitting the extraction of fewer trees per hectare but resulting in a relatively high amount of total timber extracted. It should be noted that no complete set of treatments has yet been applied. The state of the forest after more than one rotation should be tested in the existing long-term plots where selective logging may be continued as intended under the CSS (Dekker and De Graaf 2003). Most management systems for natural tropical forests today are modifications of the SMS or the CSS geared to suit the local ecological characteristics of the forest as well as the economic conditions of the region (Box 5.1).

Box 5.1

Research on forest management systems at CATIE, Costa Rica

In Central America, long-term research at CATIE (Centro Agronómico Tropical de Investigación y Enseñanza, Tropical Agriculture Research and Higher Education Center) has focused on developing technologies for sustainable management of natural forests. Researchers have recently created models to predict and simulate growth and yields of natural forests and have generated quantitative and qualitative information on ecological and economic feasibility of natural forest management systems in the region (Montagnini et al. 2002).

A financial analysis of sustainable management in a harvested forest was recently conducted at the Tirimbina Rain Forest Research Center, located in Costa Rica's Atlantic zone. The Tirimbina forest is part of a network of key sites for long-term research by CATIE on sustainable forest management in tropical America. Based on studies of economic feasibility and impacts on plant biodiversity of timber extraction, it was found that at least 30 ha with 10–15 m^3 of timber/ha should be harvested at each intervention if management is to be economically attractive and ecologically sustainable (Campos et al. 1998).

In these forests, post-harvest silvicultural treatments increased growth, especially for commercial species. Simulations using SIRENA, a growth and yield model (De Camino 1997), suggested that sustainable management can be achieved when harvesting is kept to moderate levels, when post-harvest treatments are applied to maintain an appropriate composition of commercial species, and when a cutting cycle of at least 20 years is used.

Other management systems in Latin America follow similar guidelines. It is still too early to know whether these systems are indeed both ecologically and economically sustainable. However, their chances are highly increased if they follow the criteria and indicators of sustainable management described in Section 5.4.1.

5.2.2.2 Partial Clearing Systems

Tropical Shelterwood System (TSS)

TSS was introduced in Nigeria and Ghana in the 1940s. In contrast to the MUS, the forests under management did not have enough regeneration potential, and therefore canopy openings were done several years prior to harvesting in order to promote adequate regeneration. Canopy opening by felling or poison-girdling selected trees was prescribed over a 5-year pre-harvest period. In practice, every tree not considered to have economic value was killed, resulting in severe canopy openings (65–80% of total basal area). This drastically increased light levels, leading to growth of vines and other light-demanding weeds instead of the desired regeneration of hardwood species. The system as practiced was not very effective. TSS could potentially work only in forests where light-demanding species are the desirable species and where climbers and weeds are not a big problem. In addition, the TSS is often too expensive for low-yielding forests (Dawkins and Philip 1998). The TSS was gradually abandoned in Ghana and Nigeria in the late 1970s and polycyclic methods were adopted. Variations of the TSS are sometimes utilized in tropical forests, generally as an initial treatment preceding other management operations.

Palcazu Method (Strip Cutting)

The Palcazu method, designed and applied in the Palcazu Valley of eastern Peru in the 1980s, was based on ecological observations of gap-phase dynamics of tropical forests. In natural forests, following gap formation, shade-intolerant species regenerate in the center of the gap, and more shade-tolerant species tend to occupy the borders. In the Palcazu method, the forest was cut in long narrow strips 30–40 m wide, simulating natural forest gaps to maximize utilization of the timber and to facilitate natural regeneration of trees. The strip clear-cuts were rotated through a production forest so that uncut primary forest or advanced secondary forest bordered a harvested strip. One strip was cut every year. Forest cutting was done using directional felling, cutting the trees so that they fell towards the already open area. Lianas were cut the year before felling. Extraction was done by oxen. A 40-year cycle was originally planned, to be done in parallel strips, leaving mature forest in between. Forest management using this strip cutting method was practiced on a total of 50,000 ha in an area with low hills. The project was based on an integrated use of products (sawn timber, poles, charcoal). There were two small sawmills on site to process the extracted timber. Initial evaluations of demonstration experiments using strip-cutting in the forest of the region were promising and showed that harvesting and extraction of wood could be done under the local conditions without serious environmental damage and that initial regeneration was rapid, abundant, and diverse (Hartshorn 1990).

The project was initially funded by the US Agency for International Development (USAID) to serve social objectives for communal lands of indigenous people living in the region, who were associated in a cooperative. The project suffered several difficulties after the withdrawal of the external financial and technical aid in 1989 due to the guerrilla activity in the region (Dawkins and Philip 1998). The cooperative continued the project for several more years but was faced with a variety of marketing and administrative difficulties. Costs exceeded revenues as timber prices were kept low under the national government policies at the time.

Yields were also lower than predicted by the pre-felling inventories. Silvicultural treatments were experimentally applied to the previously cut strips to test the effects on regeneration. Thinning significantly enhanced annual growth increments for stems in all regenerating categories of commercial species, with growth rates approximately twice those in the controls (Dolanc et al. 2003). However, economic sustainability of the system is still in question, as growth rates of commercial stems, even in thinned plots, were quite low (<0.3 cm/year for all categories).

5.3 Reduced Impact Logging (RIL)

Harvesting and extraction operations are the activities that generally cause the most significant impacts on the forest (Fig. 5.1). They include all the activities necessary to fell trees and remove them from the forest to the log site for loading and transport. Selective logging leads generally to a variety of short-lived and long-lived effects including changes in the light regime and forest micro-climate, erosion, soil compaction, disruption of nutrient cycling, and possibly long-term changes in tree species composition. These changes can affect the recruitment of timber species and the diversity of forest fauna. Selective logging also increases the forest's susceptibility to fire through modification of the understory micro-climate and supply of fuel (Pereira et al. 2002).

To make forest management systems more "environmentally friendly", silvicultural and management schemes have concentrated on decreasing forest damage by lowering the intensity of timber harvests and improving logging

Fig. 5.1. Manual extraction of sawn wood decreases forest damage in this communal forest of the Toncontín Agroforestry Group in La Ceiba, Honduras. The wood is sawn to boards of various sizes in a small frame sawmill in the forest. They transport the sawn wood 6 km average distance to the community, sometimes using also mules or horses. (Photo: CATIE)

practices (Bertault and Sist 1997; Sist et al. 1998). The quality of planning and the execution of harvesting and extraction are crucial in determining the state of the forest ecosystem following harvesting. The components of the reduced impact logging (RIL) approach can be adjusted to fit the specific forest conditions of each region or management unit, but they generally include (1) inventory and mapping to reduce waste during logging; (2) planning of roads, log decks, and skid trails to minimize ground disturbance; (3) vine or liana cutting 1 year prior to harvest to improve worker safety and eliminate damage to neighbors of harvested trees; (4) planned directional felling and bucking to minimize damage to future harvests and reduce waste; and (5) planned extraction to minimize equipment time during skidding. These practices may be complemented by silvicultural treatments to improve the long-term prospects for forest stand productivity.

RIL practices significantly limit damage compared to conventional logging practices, particularly at low or intermediate harvest intensities (Boxes 5.2 and 5.3).

Box 5.2

Impacts of RIL and conventional logging on forest damage in Indonesia

Sist et al. (1998) examined the impacts of conventional and reduced-impact logging (RIL) in East Kalimantan, Indonesia, by comparing pre- and post-harvest stand inventories. Felling provoked injuries in the remaining trees, mainly in their crowns. Skidding was the major source of tree mortality. RIL decreased skidding damage but failed to control felling damage. There was a positive correlation between logging intensity and the proportion of trees damaged by felling. If logging intensity was high (>8 trees/ha), the proportion of trees damaged in RIL was similar to conventional logging. They concluded that RIL techniques can reduce logging damage by 50% in comparison with conventional logging, if logging intensity were kept low or moderate (<8 trees/ha). The techniques used to extract trees should follow the specific restrictions of the RIL guidelines. For example, felling is not allowed in and around ecologically sensitive areas such as riparian strips or steep terrain. RIL also requires a higher rejection of trees by fellers when the felling direction is unpredictable. Results of more recent studies on the same forests led to new silvicultural rules to be used as part of RIL in these forests. These rules specify that it is necessary: to keep a minimum distance between tree stumps of about 40 m; to ensure only single-tree gaps; to use directional felling; and to harvest stems only in the 60- to 100-cm dbh range (Sist et al. 2003). The authors also suggest that RIL techniques should be expanded beyond silvicultural concepts, including the maintenance of other goods and services of the forest.

Box 5.3

Comparison of forest damage between RIL and conventional logging in the Amazon

In the eastern Amazon, the use of RIL resulted in about half to one-third the ground damage (roads + skid trails + log decks) when compared to conventional logging (CL). Canopy damage was also about half in RIL compared with CL treatments (Pereira et al. 2002). At this same site in Brazil, a comparison of costs and revenues for typical RIL and CL operations included estimations of productivity, harvest volume, wasted wood, and damage to the residual stand. The major conclusion of the study was that RIL was less costly and more profitable than CL under the conditions observed in the eastern Amazon site. The largest gain to RIL was provided by savings on the otherwise wasted wood. Large gains attributable to RIL technology were also observed in skidding and log deck efficiency. In addition, investment in RIL yielded an environmental dividend in terms of reduced damage to trees in the residual stand and reduction of the amount of ground area disturbed by heavy machinery (Holmes et al. 2002). The authors concluded that monetizing the value of the environmental dividend remains a major challenge in the promotion of sustainable forestry in the tropics.

RIL can also lead to economic savings. For example, in the Celos system in Surinam, the increased costs of planning with the use of RIL were found to pay off in increased efficiency (especially in skidding), reduced wastage, as well as reduced environmental damage. Savings in skidding costs with the use of RIL have also been reported in Sarawak (Higman et al. 1999). However, some studies have found higher costs in RIL than in conventional logging, due to the need for extra training, higher standings for road building, and higher costs of supervision. The fact that RIL is still of limited use throughout the tropics suggests that either it is more expensive or potential financial advantages are outweighed by other considerations (Leslie et al. 2002).

Logging impact alone is not always a good measure of the quality of forests that remain after logging (Wadsworth 1999). If better forest conservation is to result from low impact logging, it may require additional practices to those generally embodied in RIL. For example, directional felling could significantly reduce logging impacts if its purpose, in addition to ease of skidding, was also to avoid damage to immature trees. In sum, there is more to RIL than just guidelines on how to reduce logging damage. RIL should always be used in the broader context of sustainable forest management. RIL on its own may help reduce logging damage, but in isolation it will not ensure better forests.

A key aspect in sustainability of forest management is logging frequency. Logging intensity can be characterized by two aspects: static intensity (one-time logging) and dynamic intensity (frequency, i.e., cutting cycle). The regulation of frequency of timber removal (implementation of a cutting cycle) requires a management unit large enough to accommodate a reasonable cutting cycle, generally taken to be between 20 and 30 years, and at the same time large enough annual compartments so that the static harvest intensity can be kept reasonably low. As a result, sustainable management depends upon an adequate formulation of management units for which appropriate long-range management plans must be developed, specifying the manner in which logging intensity will be regulated, and the means by which productivity can be enhanced through silvicultural practices (Vincent 2002).

5.4 Ecological and Economic Feasibility of Methods of Management of Natural Tropical Forests

As seen from the description of the methods of management of tropical forests throughout the world, a variety of factors can affect their ecological, economic, and social feasibility. A key factor associated with the nature of tropical forests is the wide range of silvicultural characteristics of the desired species. Foresters need to have knowledge of the ecology and silviculture of several species in order to be able to design treatments that should be applied to suit the preferred species. For example, foresters need to know the light requirements for growth of the preferred species, so that they can apply the proper refinement and liberation techniques. There needs to be an adequate knowledge on tree species composition of the forest, both for performing pre-harvest inventories, as well as for evaluating the status of natural regeneration. Knowledge of the reproductive ecology of species is also needed; for example, if some of the harvested species are dioecious, care must be taken to ensure that other individuals of both sexes are present to ensure proper regeneration. Following initial treatments or canopy openings, additional silvicultural treatments (such as liberation and refinement) generally are needed to stimulate seedling regeneration in response to canopy opening and increased light availability.

Success of a management system lies in answering the following questions: are there enough seedlings, saplings, and advanced growth of merchantable species at time of exploitation to provide adequate stocking for the next harvest? What are the silvicultural characteristics of these species? What treatments will be necessary? What are probable growth rates and merchantable volume expectations of different species? What are the costs of the treatments? What is the cost of RIL?

In addition, in order for a management system to make sense from an economic point of view, an integrated land use policy is needed, where forest management is only part of the economic activities in a region. Forest management makes more sense economically when it is complemented by agriculture, tourism, or other activities. Socially, a successful forest management system has to provide safe employment to local people and suit their needs and preferences. For example, some successful forest management schemes are practiced by people who own forests in a communal system (explained in Chap. 7).

Given the wide variety of methods employed to manage natural forests in the tropics, efforts have been made to standardize the principles and criteria used to determine if a management system is ecologically, economically, and socially sustainable. A set of indicators has been defined to aid in the evaluation of the management systems suited to the particular conditions of each region, as explained in the next section.

5.4.1 Criteria and Indicators of Sustainable Forest Management

The International Tropical Timber Organization (ITTO) was created by treaty in 1983 with the objective of providing an effective framework for consultation among producer and consumer member countries on all aspects of the world tropical timber economy. ITTO brings together 56 member nations with interests in the trade of tropical timber and the management of tropical forests. The organization's task is to foster a tropical timber trade that simultaneously contributes to development in tropical countries and conserves the tropical forest resource on which the trade is based (www.itto.or.jp).

ITTO has pioneered the development of criteria and indicators (C & I) for the sustainable management of natural tropical forests. These are designed to assist tropical countries in assessing and reporting on compliance with forest management standards. A criterion describes a state or situation that should be met in order to comply with sustainable forest management. Seven criteria are identified as essential elements of sustainable forest management: criterion 1, *Enabling Conditions for Sustainable Forest Management*, is concerned with the general legal, economic, and institutional framework without which actions included under the other criteria could not succeed. Criteria 2 and 3 on *Forest Resource Security* and *Forest Ecosystem Health and Condition*, respectively, are concerned with the quantity, security, and quality of forest resources. The remaining four criteria deal with the various goods and services provided by the forest – *Flow of Forest Produce*, *Biological Diversity*, *Soil and Water*, and *Economic, Social, and Cultural Aspects*. The indicators have been carefully identified and formulated so that a change in any one of them would give information that is both necessary and significant in assessing progress towards sustainable forest management. Wherever possible, quantitative indicators have been suggested, but, in some instances, this is impossible or prohibitively expensive. Where this is the case, qualitative or descriptive indicators are provided.

The purpose of ITTO's C & I is to provide member countries with an improved tool for assessing changes and trends in forest conditions and management systems at the national and forest management unit levels. These indicators identify the information needed to monitor change, both in the forest itself (outcome indicators) and in the environmental and forest management systems used (input and process indicators). The information generated using these C & I in assessing the state of the forest helps policy-makers to communicate their efforts towards sustainable forest management more effectively to the public. It can also assist in developing policies and strategies for sustainable forest management, in focusing research efforts where knowledge is still deficient, and in identifying those areas that are in special need of international assistance and cooperation.

The incorporation of ITTO's C & I in the national policies of many member countries is a major achievement, but implementing them at the forest level remains an enormous challenge (ITTO 2003a). Since ITTO undertook its pioneering work in the early 1990s, several international and regional initiatives on criteria and indicators for sustainable forest management have emerged, stemming from the UN Conference on Environment and Development held in Rio de Janeiro in June 1992. These initiatives involve more than 100 countries and include the Pan-European Helsinki Process, the Montreal Process for temperate and boreal forests, the Tarapoto Proposal for the Amazon, and regional initiatives for Dry-Zone Africa, the Near East, Central America, and the African Timber Organization (ITTO 2003b). In February 1997, the UN Commission on Sustainable Development's Intergovernmental Panel on Forests endorsed the concept of criteria and indicators for sustainable forest management and called on all countries to become involved in implementing them.

5.4.2 Certification of Forest Management

Forest management certification can be seen as a way to verify the quality of the management operations by an independent third party. Forest certification was set up as an instrument to transfer costs of sound forest management from forest owners to consumers. It can function as an economic incentive to producers and consumers that commit themselves to a more responsible use of natural forests (Upton and Bass 1996).

The Forest Stewardship Council (FSC) was founded in 1993 to promote good forest management by evaluating and accrediting certifiers by encouraging the development of national and regional standards of forest management and strengthening national certification in developing countries. The FSC has declared ten principles and criteria; these are: (1) compliance with laws and FSC principles; (2) tenure use rights and responsibilities; (3) indigenous peoples' rights; (4) community relations and workers' rights; (5) benefits from the forest; (6) environmental impact; (7) management plan; (8) monitoring and assessment; (9) maintenance of high conservation value forests; and (10) plantations (Higman et al. 1999). Their principles and criteria direct attention to environmental, conservational, non-timber production, and social objectives, more than to the sustained production of timber; yet the certification scheme to provide consumers with reliable information about the source of forest products is centered on wood (Dawkins and Philip 1998). The FSC global principles and criteria are being adapted into regional and national standards worldwide in order to incorporate locally appropriate interpretations of performance standards. Once endorsed by the FSC Board, the local standards can be used by FSC-accredited certifiers when working in those regions (Higman et al. 1999).

However, certification of forest management does not necessarily guarantee sustainability. Rather, it provides effective and credible independent proof that the forest is being well managed, and presumably this will result in sustainability (Higman et al. 1999).

Although the cost of certification may not be expensive on a per hectare basis (Upton and Bass 1996), financing a certification scheme can be difficult when cash flows are precarious. An additional difficulty lies in finding staff with sufficient training to carry out certified forest management operations. If prices for certified wood do not compensate the cost of certification, there is little incentive to certify. However, there are non-financial benefits to certification. The certification process often serves as an interface between research, forest policy, and management. In Costa Rica and Guyana, for example, certification has served to link forestry research and policy. In these two countries, the process of choosing a certification scheme and developing a national certification standard and forestry policy was built upon existing research (Louman et al. 2002).

The C & I for sustainable forest management can also be used to define the criteria for certification of forest products. Certified products are more appealing to consumers concerned about the environment and often sell for better prices. In addition, some forest owners often pride themselves on practicing sustainable forest management and certification of forest products is a standard way to demonstrate sustainability.

5.4.3
Obstacles to Sustainable Forest Management

The idea that sustainable forest management can be an effective conservation tool rests, in part, on the premise that it can stabilize wood production in a given area. In principle, this would lead to conservation by maintaining forest cover and reducing pressures on other primary forests. However, this has rarely happened in practice (Bowles et al. 1998). There are steep hurdles facing broad adoption of investments in SFM. For one thing, such investments are almost always financially unattractive. Reaping a one-time harvest of ancient trees today is simply more profitable than managing for future harvests (Reid and Rice 1997).

Most countries with tropical forests do not have the capability to counter such financial incentives. Tropical countries are often feeding grounds for foreign logging corporations. Many developing countries have severe economic problems such as high inflation, unemployment, and foreign debt, and as a result are usually willing to attract investment (see Chap. 4). Many of these countries have weak environmental and social laws or little enforcement capabilities and are thus highly attractive to foreign loggers. Bribery and corruption can be severe problems because forest resources are often controlled by a few powerful individuals (Laurance 2000).

Just because there are serious obstacles to SFM does not mean that researchers and practitioners should give up the idea. The stakes are too great. Rather, it means that the effort to promote SFM must be redoubled and refined. There is already evidence that reduced impact logging, an important component of SFM, is economically attractive when looking toward the long-term profitability of the forested area. The problem arises when those with the authority over the resource have no responsibility for the future of that resource. Resolving the problem of exploitation of tropical forests means that *authority* over the land and the forest upon it must be more closely linked to the *responsibility* for the future of that resource.

5.5
Management of Secondary Forests

The term "primary forest" is used to designate a forest that has fully recovered formany past disturbance, and the structure and function resemble a forest that has never been cut. The term "secondary forest" is used for those forests that have recently regenerated following a natural (hurricanes, landslides) or human-induced (logging, clear-cutting, and replacement by agriculture or other land use) disturbance. Secondary forests in the tropics occupy larger and larger areas as the area of primary forest decreases and agricultural areas are abandoned due to unsustainable practices. For example, in Central America, secondary forests are rapidly growing on abandoned pasture lands (Kaimowitz 1996).

The area of secondary forests has been estimated to increase at a rate of about 1 million ha/year (Achard et al. 2002). The type and rate of formation of secondary forests varies from region to region. There are two broad categories of secondary forest. One is residual forest that has been cut over once or more in the past 60–80 years. Because they have never been completely felled, they retain some of their former characteristics. The other type, called fallow or "volunteer", is forest that has regrown after being clear-cut for agriculture, pasture, timber extraction, or some other use. Because volunteer forests are composed largely of pioneer species, they lack both the structure and the composition of a mature forest. About 55% of secondary forests in the tropics as a whole are cutover forests and 45% are fallow (volunteer) (Wadsworth 1997).

Most secondary forests that developed after selective timber extraction are located in tropical Asia (47%), followed by tropical America (32%) and tropical Africa (21%) (Brown and Lugo 1990). The structure and composition of these secondary forests vary according to forest age, site fertility, previous land use, and distance from seed sources. The economic potential of each secondary forest for management for timber has to be determined with inventories that take into consideration the marketability of possible products. In general, secondary forests that originated from abandoned agricultural lands are closer to centers of human population, and therefore access and markets are not a problem. Most of the species growing as volunteers are heliophilous (light-demanding) and therefore likely to respond positively to silvicultural treatments such as thinning and liberation. A potential disadvantage of secondary forests as timber sources is the relatively lower market value of the timber species present, especially in comparison with primary forests. However, the markets change as the supply of more valuable timbers diminishes. Many secondary forests are good sources of fuelwood, non-timber forest products, pulp, and timber for local use.

In addition, tropical farmers have long depended on secondary forest fallows to restore productivity to land worn out by cultivation. When a secondary forest replaces a crop or pasture, the production of biomass by the vegetation and the cooler soil temperatures under the forest canopy contribute to the addition of organic matter to the soil. Typically, in most tropical humid areas, fallow periods of 5–15 years are required for soils to recover organic matter so that soils can be farmed again (Van Wambecke 1992). The type of secondary vegetation and the species composition influence the rate of recovery of soil fertility and the specific nutrient inputs to the soil. Farmers often enrich the fallow with fruit trees or other useful species, thus making use of the fallow and sometimes even accelerating the recovery of soil fertility.

5.5.1 Techniques for Management of Secondary Forests

The techniques used in management of secondary forests are generally similar to those used in management of primary forests, consisting of some type of selective management, generally followed by silvicultural treatments. Forest enrichment techniques are also often used as a way to increase the biological and economic value of secondary forests (Montagnini et al. 1997). Forest enrichment techniques are discussed in Chapter 6.

In refining secondary tropical forests, foresters select crop trees according to their prospective marketability, present size relative to maturity, form, freedom from injuries, and apparent health. If there is advanced regeneration of shade-tolerant, late-successional species (preferred crop trees) and the canopy is dominated by light-demanding mid-successional species, the only way to successfully regenerate another secondary stand is to remove all commercial volume at once in a monocyclic system (Guariguata 2000). In contrast, if a stand has plenty of shade-tolerant poles of good commercial value, a polycyclic system may be possible. Finegan (1992) suggested rotations of 15–25 years for neotropical secondary forests with light-demanding commercial species, depending on species composition and production goals. Silvicultural treatments should be applied when a closed canopy of longer-lived species has been formed. Under a monocyclic system, treatments will focus on liberating light-demanding crop trees from competition in the canopy, while under a polycyclic system, both light-demanding canopy species and advanced regeneration of shade-tolerant species are considered for liberation (Kammesheidt 2002; Box 5.4).

Table 5.2. Annual median diameter increment (in cm) of crop trees of the species studied, 2 years after thinning in a young secondary forest in the Caribbean lowlands of Costa Rica. (Data from Guariguata 1999)

Species	Control	Thinned
Laetia procera	0.4	1.0
Tapirira guianensis	0.5	1.4
Simarouba amara	1.7	1.8
Vochysia ferruginea	0.7	1.8
All species	0.7	1.2

Fig. 5.2. A *Cecropia* tree in a secondary forest on the Pacific coast of Costa Rica (note a sloth hanging from a branch near the center). Several species of *Cecropia* are characteristic of early stages of succession in secondary forests in Latin America. (Photo: F. Montagnini)

Box 5.4

Research on management of secondary forests by CATIE

Despite the large body of ecological information on secondary forest succession in Central America, few forestry-based experiments have investigated how secondary forests react to management practices. CATIE researchers are characterizing secondary forest structures and floristics and developing guidelines for sustainable management in Costa Rica and Nicaragua (Current et al. 1998; Guariguata 1999). In Costa Rica, the area of secondary forests already exceeds the area of primary forest legally available for production, while in Puerto Rico nearly all forest cover is classified as secondary forest (Kammesheidt 2002; Fig. 5.2).

In Costa Rica, CATIE has investigated the effects of silvicultural practices such as liberation thinning, whole-canopy removal, and substrate preparation techniques on stand dynamics and regeneration of secondary forests in order to provide guidelines for sustainable management of timber. In the Atlantic lowlands of Costa Rica, short-term growth responses in individuals of four commercial species (*Laetia procera*, *Simarouba amara*, *Tapirira guianensis*, and *Vochysia ferruginea*) were evaluated following liberation thinning in a young secondary forest. Liberation thinning significantly increased the diameter growth of future crop trees with respect to unmanipulated counterparts (Table 5.2). The study concluded that young stands in the region may be attractive systems for simple silvicultural manipulations due to rapid growth responsiveness, facilitated by manageable tree size (Guariguata 1999).

The type of intervention needed for management will significantly vary according to the status of secondary forests. In tropical Asia, five major categories can be found: post-extraction secondary forests, post-fire secondary forests, swidden fallow secondary forests, secondary forest gardens, and rehabilitated secondary forests (Chokkalingam et al. 2001). An understanding of where each particular forest is situated in a continuum of disturbances and regeneration stages can help guide management of secondary forests.

5.6 Management for Non-Timber Forest Products (NTFPs)

As seen in Chapter 1, tropical forests provide a wealth of plant and animal products and a variety of environmental services. Management of forest resources must be viewed in the context of the surrounding land and natural resources. Likewise, management for timber cannot be completely separated from management for NTFPs. In fact, timber extraction often affects populations of NTFPs, including plants and animals.

Designing systems for diversified forest management involves studies on the ecology and management of non-timber species, including trees, herbs, and palms used locally or regionally for medicine, insecticides, ornamental purposes, craftwork, and construction (Fig. 5.3). Sustainable management for the use of these resources is based on studies on the capacity of each species to supply the desired products in a sustainable manner. Silvicultural guidelines for management are developed for each species. Development of silvicultural systems that include sustainable management plans for NTFPs requires knowledge of the biology and uses of the species, including the population structure and estimated amount of harvestable product.

In many cases, traditional extraction may lead to the exhaustion of the NTFP resource (Marmillod et al. 1998). For example, Brazil nuts (*Bertholletia*

Fig. 5.3. Palms are important non-timber forest products that are extracted from forests for a variety of uses. Some species of *Chamaedorea* palm, known as xate, are extracted from forests in Guatemala and other countries of Central America to export to the USA for ornamental uses. (Photo: F. Montagnini)

excelsa) are a classic NTFP and are the only internationally traded seed crop collected exclusively from natural forests (Fig. 5.4). A recent analysis of 23 populations of the Brazil nut tree across the Brazilian, Peruvian, and Bolivian Amazon has shown that populations subjected to high levels of harvest lacked juvenile trees; only populations with a light history of exploitation contained large numbers of juvenile trees (Peres et al. 2003). Without proper management, intensively harvested populations may succumb to a process of senescence and demographic collapse, threatening this cornerstone of the Amazonian extractive economy. Another example is the ornamental plant *Zamia skinneri* in Central America, whose excessive extraction is leading to very low populations in natural forests (Box 5.5).

Fig. 5.4. A Brazil nut tree (*Bertholletia excelsa*) in the delta of the Amazon River in Pará, Brazil. (Photo: F. Montagnini)

Box 5.5

The extraction of *Zamia skinneri* from Central American forests led to its inclusion in Appendix II of CITES

Zamia skinneri, a 2.5-m palm-like cycad (Cycadaceae family), is currently included in Appendix II of CITES (Convention on International Trade in Endangered Species) (Robles et al. 1997; Maiocco 1998). Only commercial exports of seeds are allowed for species included in Appendix II. *Zamia skinneri* used to be extracted from natural forests throughout all its natural range in Costa Rica, Nicaragua, and Panama to be sold as an ornamental plant, in both national and international markets. Several species of *Zamia* grow naturally from the southern USA to Bolivia. The Central American species of *Zamia* live in the understory of tropical humid forests. *Zamia skinneri* grows in an altitudinal range from 20–1,100 m. It only grows in forests and cannot be grown in open plantations because it cannot tolerate direct exposure to the sun. *Zamia skinneri* is an ancient species, often considered a living fossil. In the Dominican Republic, archaeological records show that *Zamia* spp. was used by indigenous tribes about 1400 years b.c. The Banwari people of the Dominican Republic used to cook and eat the stems of this plant, in a manner that is similar to the use of cassava or manioc roots. Its value as food contributed to the protection of the forests in which it grows (Maiocco 1998). In present times interest in *Zamia skinneri* has grown because of its use as an ornamental. Much of the demand comes from the US. Due to its endangered status and high commercial value, researchers at CATIE have been studying its ecological requirements in order to design more sustainable manners of extraction from natural forests or cultivation in the appropriate environment (Robles et al. 1997; Maiocco 1998; Marmillod et al. 1998).

Extraction of NTFPs is not a priori more sustainable than timber extraction. On the contrary, NTFPs are equally threatened by overexploitation and abuse. In addition, changes in the social structure or living conditions of local people may lead to abandonment of NTFP extraction. In most cases, NTFP extraction is carried out by relatively poor sectors of the population. If their economic conditions change they may choose other less laborious and more profitable activities (Bruenig 1996).

Market forces also tend to impede sustainable management of NTFPs. Once an NTFP enters the cash economy, the usual cycle – establishing a market, rising demand, more intensive harvesting, collapse of the price, and finally substitution – tends to develop (Dawkins and Philip 1998). As demand increases there is a trend away from wild collecting and towards commercial domestication and cultivation. For example, wild-collected rubber from the Amazon could not compete with rubber from plantations of rubber trees in Asia. Even in such poverty-stricken regions as Amazonia, the long-term trend of social and economic evolution towards improved living conditions may make the collection of NTFP in natural rainforests less attractive.

Management for NTFPs can only be viable in the context of other land uses and economic activities for human populations. For example, there are reported cases where communities have organized themselves to crop communal land, extract timber, manage the forests for NTFPs, and reserve some forest areas for ecotourism (Montagnini et al. 2002).

5.7 Is Forest Management Compatible with Conservation of Biodiversity?

Many tropical forests have been subject to low-intensity human management for centuries: this management generally preserves environmental functions and species diversity (Gómez-Pompa 1991). Because of increased awareness of the importance of sustainability and the preservation of biodiversity, many tropical countries have recently changed forest management regulations to make them compatible with the principles of sustained yield and biodiversity preservation (Boyle and Sayer 1995). The ITTO and Forest Stewardship Council principles, criteria, and indicators for SFM give special emphasis to conservation of biodiversity. Specific guidelines are needed to cover the vast array of forest conditions that are present in each situation: species, soils, environmental constraints, markets, and other factors. The management guidelines should be adjusted to suit the scale and objectives of management in each particular case. In addition, methods are needed to evaluate ecological indicators that can serve to verify effects of management on long-term forest productivity and maintenance of biodiversity (Lowe 1995).

Operational definitions of diversity are required to provide practical and consistent frameworks for measuring and monitoring biodiversity. Typically, four different levels of diversity have been considered for measuring and monitoring biodiversity: genetic, species, ecosystems, and landscape levels (Boyle and Sayer 1995). Species diversity is conceptually the simplest of the four levels of organization of biodiversity, both because many species are visually distinct and because species extinction is an emotionally dramatic event. Practical measures of biodiversity tend to focus on the species level. Studies of genetic diversity have been done for several tree species of commercial value, and in some cases they have led to regulations about species conservation (e.g., a ban on logging mahoganies in Costa Rica; Navarro et al. 2002).

There are many published studies of the effects of forest management on biodiversity. For example in a forest reserve in Misiones, Argentina, a "Uniform Spacing" (US) method of forest harvest was used, where trees were selected for extraction or marked for retention according to scarcity, horizontal distribution, and quality as seed trees (Fig. 5.5). The system is called Uniform Spacing because the trees remaining after its application tend to be uniformly spaced. Generally the cutting intensity in the US method is about half that of conventional logging methods. In this case, no silvicultural treatments were applied following timber harvest. Forest regeneration following timber harvest was compared between the US method and conventional logging methods to see the effects of the different harvest techniques on forest biodiversity. Three years after harvesting, the forest cut by uniform spacing had the highest total density of seedlings for commercial and non-commercial species, and also exhibited

Fig. 5.5. *Ocotea puberula* trees are generally extracted from forests in Misiones, Argentina, due to their high timber value. However, in the Guaraní Forest Reserve, when cutting was done using the Uniform Spacing method, this individual was left standing, while a nearby tree of the same species was cut (note remaining stump in *foreground*). (Photo: F. Montagnini)

higher diversity of understory plants compared with forest cut using conventional logging methods (Montagnini et al. 1998). In the US system the resulting forest had higher variability of microenvironments which led to the establishment of a greater variety of species. In contrast, in the conventional cutting method the intensity or extraction was higher, resulting in greater canopy openings, which led to the establishment of more bamboo, ferns, and grasses. Other studies in Costa Rica also report on changes in forest composition following post-harvest silvicultural treatments (Box 5.6).

Box 5.6

Changes in species composition following silvicultural treatment in Costa Rica

In Costa Rica, results of a silvicultural experiment carried out by CATIE in a primary forest focused on the effects of logging and post-harvest silvicultural treatments on forest species richness and composition during the first 6–7 years following logging and 5 years following the application of silvicultural treatments (liberation and refinement) (Montagnini et al. 2001). The forest studied exhibited marked compositional variation in relation to a topographical gradient after the implementation of the experiment; such β- or "ecosystem diversity" should be taken into account in evaluations of the effect of forest management on plant biodiversity. In this forest, post-harvest silvicultural treatments caused an immediate reduction in species richness in individuals greater than 10 cm dbh. This was due to the chance elimination of species represented by one or a small number of individuals in the plots. It was concluded that post-harvest silvicultural treatments may affect overall species composition by favoring commercial species. However, no changes of species richness or composition were evident in the forest understory (individuals between 2.5–9.9 cm dbh). The direct felling and extraction of timber caused the only detectable changes in understory plant biodiversity and these changes were found only in the localized areas disturbed by these management operations.

Another study in Latin America reports that conventional low-intensity mahogany harvesting in lowland Bolivia has only a relatively mild physical effect on the forest. However, it is doubtful if this would hold true for the much more intensive harvesting characteristics of the eastern Amazon or the dipterocarp forests of Southeast Asia (Pearce et al. 2003).

In the forests of Central Africa, Hall et al. (2003) compared forest structure and tree species composition between unlogged forest and forests that had been subjected to highly selective logging 6 months and 18 years prior to the study. While there was little difference in tree species composition and diversity between treatments, stem densities were significantly higher in the unlogged forests than those in the forest sampled after 18 years since logging. In the logged forest there was insufficient recruitment of the principal timber species *Entandrophragma* spp. (African mahogany). However, many other quality timber species remained after selective logging, making logging still attractive long after elimination of the preferred species.

Some species function as indicator species, meaning that their presence is indicative of high biodiversity and a well-functioning ecosystem (Lindenmayer et al. 2000). Such ecosystem functions can include stand structural complexity, plant species composition, connectivity, and heterogeneity. Carefully designed studies are needed to test the relationships between the presence and abundance of potential indicator species and the maintenance of ecosystem processes in forests.

In conclusion, SFM techniques can contribute to maintaining biodiversity in tropical forests managed for timber production. Application of silvicultural treatments following harvest may immediately reduce species richness by eliminating species represented by few individuals. The effects of forest management operations on plant biodiversity depend on the nature of the operation. Studies are clearly needed to assess the long-term effects of forest management on plant and animal biodiversity. Any systems designed to restore degraded ecosystems should be focused on recovering at least part of the lost biodiversity (Montagnini 2001).

5.7.1 Effects of Forest Management on Wildlife

Wildlife is affected by logging in many ways. The direct effects of timber extraction on wildlife depend on the intensity and frequency of logging and the species involved. Some species, such as certain insectivorous birds, can disappear completely after a single intervention (Bennett 2000). Many species such as primates and hornbills decline in numbers, whereas populations of other species such as browsing ungulates can increase due to the rapid growth of browse as the canopy is opened. Secondary effects of logging can also include increases in hunting due to the opening of roads and increases in human populations depending on bush meat as a protein source.

Particularly important to the long-term survival of the forest are any impacts that logging can have on some of the mutualisms that may exist between plants and animals. Forest animals may act as seed dispersers, and if their populations are altered due to logging or increased human intrusion, the reproduction and survival of tree species may be endangered. An exam-

Table 5.3. Number of sightings of mammalian fauna at the study sites in the Caribbean lowlands of Costa Rica over a 6-month period (Guariguata et al. 2000)

Species	Tirimbina No. (no./ha)	La Selva No. (no./ha)
Alouatta palliata (howler monkey)	4 (0.06)	8 (0.13)
Ateles geoffroyi (spider monkey)	2 (0.03)	15 (0.24)
Cebus capuchinus (white-faced monkey)	7 (0.10)	7 (0.11)
Dasyprocta punctata (agouti)	–	2 (0.03)
Puma concolor (puma)	–	2 (0.03)
Tamandua mexicana (tamandua)	–	2 (0.03)
Tayassu tajacu (collared peccary)	–	4 (0.07)
Sciurus variegatoides (squirrel)	1 (0.01)	2 (0.03)
Total	14 (0.19)	32 (0.52)

ple showing the loss of fauna from managed forests and some consequences for long-term forest survival is given in Box 5.7.

Box 5.7

Effects of logging on loss of fauna from Costa Rican forests
Many studies have examined the effects of selective logging on animal populations. In Costa Rica, the effects of faunal loss on the dispersal, predation, and survival of seeds and seedlings were recently studied in two selectively logged forests with differing levels of protection (Guariguata et al. 2000). La Selva Biological Station, owned and operated by the Organization for Tropical Studies (OTS), is protected from hunting and connected to a national park, whereas the Tirimbina forest is unprotected and not connected to a park. Seed dispersal rates by mammals were highest in the protected site. Seed survival was also higher at La Selva. The low rate of seed dispersal and survival at Tirimbina is probably related to altered mammal community composition as a result of hunting pressure and loss of habitat connectivity (Table 5.3).

Even if forests are logged with minimal stand and soil disturbance, sustained recruitment of mammal-dispersed timber species appears less likely if loss of habitat connectivity and excessive hunting pressure are combined. Production forests adjacent to parks and conservation areas may be more likely to maintain a wider spectrum of viable populations of plants and animals than forests in which logging is permitted throughout.

As seen in Section 5.4.1, ITTO standards also include biodiversity, although there is no specific mention of management practices that consider faunal diversity. Likewise, conservation of wildlife is not specifically included in the Forest Stewardship Council Principles and Criteria for SFM. The concept of eco-agriculture, or managing for diversity as well as for production (McNeely and Scherr 2003), could be extended to forestry with the term eco-forestry, meaning managing forests for biodiversity as well as for production. In such cases, consideration should be taken of the total biodiversity of the forest, including wildlife. This may lead to increases in the costs of managing and monitoring, but may be essential for the long-term survival of forest ecosystem functions.

5.8 Reserves

Deforestation, whatever its causes and motivation, is the most powerful direct threat to forest biodiversity. Some support the view that the conservation of biodiversity requires halting deforestation and keeping commercial timber production out of forests (Leslie et al. 2002). This is the principle underlying the creation of totally protected areas (TPAs). However, few countries are willing or able to place all of their natural forests in TPAs. Most countries, under present economic conditions, have no choice but to continue encouraging the harvesting of timber growing in their natural forests.

Other opinions hold that biodiversity can also be conserved in production forests under sustainable management regimes; total conservation should be the first strategy and SFM is only preferred in cases when the alternative land use is wholesale logging or dramatic land use change; in some cases it is better to allow one-time conventional logging, followed by complete protection of the area (Reid and Rice 1997). However, this seems to be quite impractical given that there is no guarantee that the logged area can be effectively protected. In addition, regeneration following conventional logging may be good in some types of forests but not in all types (McRae 1997).

In order to maximize the diversity of species that are conserved, it is essential to address biodiversity conservation at the landscape level. Some species require large areas to conserve a viable population. The question of scale is important when considering the impacts of disturbance on forest biodiversity and when designing strategies for biodiversity conservation. There are different measures of biodiversity on different scales. For tropical forests greater than 10^6 km^2 in size of Africa, Asia, and the Americas, overall or "gamma diversity" (or diversity found across large geographical regions) varies from perhaps 30,000–120,000 species of flowering plants. Smaller forest plots ranging from 0.001–0.01 km^2 in area contain from 30–300 tree species ("alpha" or "within-habitat diversity"). Less information is available on "beta" or "between habitat diversity", which describes how species composition varies from one area to another. How diversity varies among plots of similar sizes in different forests and with distance among plots is a question relevant to the design of protected areas. For example, Condit et al. (2002) found that regions in Panama and the western Amazon that are 10^4 km^2 in area support 3,500–5,000 tree and shrub species, yet at smaller scales (0.01 km^2), the western Amazon forests support two to ten times as many species as do Panamanian forests. This raises the question, is it more worthwhile to preserve an Amazon forest than a Panamanian forest of similar size? Numbers may not be the only factor that matter when considering the preservation of species; the answer may depend on which species are present – their rarity, scarcity, and value (known and potential but unknown) to humans.

There are difficult issues relating to the size and spatial arrangement of protected and managed areas. The general conclusion is that protected areas should be as large as possible, but that ultimately their value as refuges depends more on their integrity and optimal distribution across the landscape than on their absolute size (Boyle and Sayer 1995). Several studies have shown that forest fragment sizes and degree of isolation are the prime determinants of species loss (Boyle and Sayer 1995). For example, Laurance et al.

(1997) reported that rain forest fragments in central Amazonia experienced a dramatic loss of above-ground tree biomass that was not offset by recruitment of new trees. These losses were largest within 100 m of fragment edges, where tree mortality was sharply increased by microclimatic changes and elevated wind turbulence. Permanent study plots within 100 m of edges had lost up to 30% of their biomass in the first 10–17 years after fragmentation. Habitat fragmentation affects the ecology of tropical rain forests by altering the diversity and composition of fragment biotas and changing ecological processes like nutrient cycling and pollination (Laurance et al. 1997). Therefore a strong case can be made for the maintenance of corridors of undisturbed forest linking refuge areas.

The conservation of species in isolated fragments of TPAs will be enhanced if these areas are surrounded by areas of modified, but biologically diverse, buffer zones, transition zones, and corridors (Boyle and Sayer 1995). ITTO's Guidelines for the Conservation of Biological Diversity in Tropical Production Forests suggest that there will be some degree of biodiversity loss in tropical production forests that would be mitigated by a comprehensive and integrated TPA network. The function of production forests in biodiversity conservation would be to allow the persistence of a large portion of the original biodiversity within a buffer zone around the TPAs, and to provide corridors that allow the free flow of genetic material among the forested areas (Leslie et al. 2002).

Some innovative plans, such as the UNESCO Man and the Biosphere (MAB) program's worldwide network of biosphere reserves, have begun considering the needs of local people by incorporating biophysical and socioeconomic factors into the management plans of protected areas (Khasa and Dancik 1997). A typical UNESCO MAB Reserve includes a core area under complete protection surrounded by buffer zones where a variety of human activities are possible. These include SFM, agroforestry, ecotourism, and even human settlements in the outer limits of the buffer zones. Similarly, extractive reserves, in which local people are responsible for forest management, have been developed for timber products in Quintana Roo, Mexico, and for NTFP extraction (mainly rubber and Brazil nuts) in Brazil.

5.8.1 Setting Priorities

The number of species threatened with extinction far outstrips available conservation resources; therefore at some point priorities must be set. Biodiversity "hotspots" have been defined as places in the world where there are high concentrations of endemic species that are undergoing exceptional loss of habitat (Myers et al. 2000). As many as 44% of all species of vascular plants and 35% of all species of four vertebrate groups are confined to 25 hotspots comprising only 1.4% of the land surface of the Earth. Should we concentrate our efforts on protecting these "hotspots?" This is an example of how knowledge of the distribution of biodiversity can be used in designing an effective strategy for conservation and management. The hotspots strategy does not exclude other areas from urgent conservation in accord with alternative criteria. For example, patterns of speciation should also be considered when determining conservation priorities (Myers 2003). There is a large degree of overlap between the biodiversity hotspots and other internationally recognized initiatives for conservation, such as the IUCN/WWF International Centers of Plant Diversity and Endemism and the endangered eco-regions of the WWF/US Global 200 List. The hotspots thesis has the potential to reduce the mass extinction under way by about one-third and has been considered as the most important contribution to conservation in the last century (Myers 2003).

5.9 Conclusion

Systems for sustainable management of tropical forests take into consideration those aspects of tropical forest ecology that are essential for the future existence of the forest:

- biodiversity, not only for the intangible benefits of biodiversity, but also because the diverse mutualisms that depend upon the survival of each species are essential for forest reproduction;
- nutrient retention and recycling mechanisms of the forest, especially the organic material on top of soils or near their surface.

The solution to maintaining diversity and associated mutualisms, maintaining nutrient cycling, and maintaining soil organic matter so essential to all functions of tropical ecosystems may lie not in any one management measure, but rather in a whole suite of measures, which will vary from country to country and situation to situation (Vanclay 1992). Furthermore, management to sustain the forest resource cannot be based upon ecological factors alone, but also must take into consideration social, political, economic, and cultural factors. These factors will be discussed in Chapter 7.

Plantations and Agroforestry Systems 6

6.1 Introduction

As tropical forests diminish worldwide, more tropical plantations and agroforestry systems are established. Because of their relatively high yields, tropical and subtropical plantations have the potential to make substantial contributions to world timber production (Evans 1992, 1999; Wadsworth 1997). Tree plantations are also a source of cash, savings, and insurance for local farmers (Chambers and Leach 1990). Over the last decade, reforestation efforts in the humid tropics have grown in response to the increase in abandoned and degraded lands. Rural farmers often respond positively to government reforestation incentives (Evans 1999), dedicating some portion of their farm to the establishment of plantations with species recommended by local technical personnel. In addition to providing timber products, tropical plantations have a function in combating desertification, protecting soil and water resources, rehabilitating degraded lands, providing rural employment, and absorbing carbon to offset carbon emissions (Montagnini and Porras 1998; Evans 1999; Keenan et al. 1999; Sedjo 1999; Whitmore 1999).

The challenge is to plan and manage plantations so as to optimize productive, environmental, and social benefits. In this chapter we discuss the importance of tropical plantations and agroforestry systems in fulfilling these environmental and socio-economic functions. We emphasize management considerations that must be taken into account when designing these systems so that they can better fulfill these multiple objectives.

6.2 Plantation Forestry: Alternative to Supplying the World's Timber Demand?

Forest plantations are defined as forest stands established by planting and/or seeding in the process of afforestation (planting trees in areas where there have never been trees before) or reforestation (replanting in areas previously supporting natural forest). According to the FAO definition, forest plantations consist of introduced or indigenous species which meet a minimum area requirement of 0.5 ha; tree crown cover of at least 10% of the land (i.e., even a rather sparse stand qualifies as a plantation); and total height of adult trees above 5 m (FAO 2000a). In some instances, terms such as "human-made forest" or "artificial forest" are considered synonyms for forest plantations.

Forest plantations covered 187 million ha worldwide in 2000 (FAO 2001 a). This represents a significant increase from the 1995 estimate of 124 million - ha. The reported new annual planting rate is 4.5 million ha globally, with Asia and South America accounting for 89% of new plantations (Table 6.1). These FAO statistics reflect not only increases in the rates of plantation establishment, but also the fact that plantations that had not been considered in previous reports, such as those of rubber trees, are included in the 2000 plantation data (Carle et al. 2002). Part of the plantation area statistics also include plantations that have been established with the purpose of rehabilitating degraded lands. Globally, half of forest plantations are for industrial use (timber and fiber), one-quarter are for non-industrial use (home or farm construction, local consumption of fuelwood and charcoal, poles), and one-quarter are for unspecified uses (Table 6.1).

During the past decade, while natural forest areas have continued to decline at the global level, forest plantations have increased in both tropical (+20 million ha) and temperate (+12 million ha) regions of the world (FAO 2001 a). In 1995, tropical and subtropical plantations comprised 45% of the global net forest plantation area. The total area of tropical and subtropical plantations was between 40 and 50 million ha (Evans 1999). Hardwoods covered 32.3 million ha, 57% of all plantations in the tropics and subtropics, and 25% of the global area (Varmola and Carle 2002). In the last decade, the rate of conversion of natural forest to plantation forest in tropical regions was about equal to the increase in forest cover, resulting from natural re-establish-

Table 6.1. Forest plantation areas by region and purpose for the year 2000. (FAO 2001 a)

Region	Total area (000 ha)	Annual rate (000 ha/year)	Industrial	Non-industrial	Unspecified
Africa	8,036	194	3,392	3,273	1,371
Asia	115,847	3,500	58,803	43,662	13,381
Europe	32,015	5	569	15	31,341
North and Central America	17,533	234	16,775	471	287
Oceania	3,201	8	189	24	2,987
South America	10,455	509	9,446	1,004	6
World total	187,086	4,493	89,175	48,449	49,463

ment (i.e., formation of secondary forest through forest successional processes) (Carnus et al. 2003).

As concerns about the status of natural forests grow and as the amount of protected areas and other large areas of forest unavailable for wood supply increases, plantations are increasingly expected to provide substitutes for products from natural forests (FAO 2000a). This is particularly true in Asia and the Pacific area, where it is estimated that more than half the natural forests are not available for wood harvest because they are inaccessible or uneconomic to exploit. Of this unavailable forest, it is estimated that about 38% is legally reserved. In addition, logging bans have been imposed on roughly 10 million ha of natural forest. The reasons for these bans vary but are related to deforestation and forest degradation in Thailand, the Philippines, and China, and to conservation requirements in Sri Lanka and New Zealand (FAO 2001b).

An analysis of the global supply and demand for industrial roundwood shows that although demand is expected to increase by about 25% between 1994 and 2010, the supply of roundwood and fiber (from both plantations and natural forests) should more or less expand to keep pace or increase slightly (Varmola and Carle 2002). Supply and demand are not, however, expected to be in balance in all regions and some classes of industrial roundwood, notably logs of high-value hardwood species from natural forest in the tropics, are expected to be in short supply due to constraints on harvesting and a shrinking resource. This is where plantation forestry is expected to play a significant role, compensating at least in part for the lack of supply from natural forests.

The potential for forest plantations to meet the demand for wood and fiber is increasing. Although plantations amount to only 5% of global forest cover, in the year 2000 plantations supplied about 35% of global roundwood. This figure is anticipated to increase to 44% by 2020, with the greatest proportional increase in Asia, doubling its industrial roundwood production. In some countries forest plantation production already contributes the majority of industrial wood supply (FAO 2000a). In China and Vietnam, the importance of plantations will increase as planted resources mature. In Sri Lanka, India, and elsewhere in the tropics, trees outside the forest are playing a critical role in roundwood and woodfuel supply (FAO 2001b).

Forest plantations of a wide range of species, including the valuable luxury hardwoods such as teak (*Tectona grandis*), mahogany (*Swietenia macrophylla*), and rosewood (*Dalbergia* spp.), have been established to meet anticipated shortages of log supplies from natural forests in the future. However, there is uncertainty about the actual extent of these plantations, their productivity, and thus their ability to supply the increasing demand. While it is clear that plantations will have an increasingly significant role in substituting products from natural forests, the impact will be felt on a case-by-case basis as governments and investors determine where and how plantations can be technically, economically, and socially feasible as well as environmentally friendly (FAO 2000b). In the near future, plantations in Asia and the Pacific area can reduce but not replace harvests from natural forests. It is likely that the current pace of industrial plantation development will barely keep pace with losses of natural forest from deforestation and the creation of protected areas. While it is theoretically possible, actual plantation development is currently not sufficient in Asia to offset both growing consumption and declining harvest from natural forests (FAO 2001b).

6.2.1 Plantation Productivity

Arguments for tropical and subtropical plantations are often based on their higher productivity in comparison with plantations in temperate regions. For example, average conifer yields are 5.3 tons ha^{-1} $year^{-1}$ in the temperate zone versus 12.6 in the tropics; for broadleaf species, the corresponding averages are 5.1 and 13.1 tons ha^{-1} $year^{-1}$ (Wadsworth 1997). Tropical and subtropical plantations are also more productive than natural forests growing in the same

Table 6.2. Mean annual increments in volume (MAI) (m^3 ha^{-1} $year^{-1}$) for several tree species grown in tropical plantations worldwide. (Adapted from FAO 2000b)

Species	MAI
Eucalyptus	
E. deglupta	14–50
E. camaldulensis	15–30
E. urophylla	20–60
E. robusta	10–40
Pinus	
P. caribaea var. *caribaea*	10–28
P. caribaea var. *hondurensis*	20–50
Other species	
Gmelina arborea	12–50
Swietenia macrophylla	7–11
Tectona grandis	6–18
Casuarina equisetifolia	6–20
Cupressus lusitanica	8–40
Cordia alliodora	10–20
Leucaena leucocephala	30–55
Acacia auriculiformis	6–20
Acacia mearnsii	14–25
Terminalia superba	10–14
Terminalia ivorensis	8–17
Dalbergia sissoo	5–8

region. For example, natural forest production in Uganda was 0.7 m^3 $ha^{-1}year^{-1}$, while nearby *Eucalyptus* plantations produced more than 40 m^3 $ha^{-1}year^{-1}$ (Laurie 1962; cited by Wadsworth 1997). In many parts of the tropics plantations yield several times the quantity of wood of most natural tropical forests and most forest in temperate regions (Evans 1992). This is true for plantations where species match the sites and when plantations are given the proper management, including genetic selection, as discussed below. Fertilization and pesticide application also contribute to the high production of plantations in comparison with natural forests.

On average, *Eucalyptus* and *Pinus* species, which dominate industrial plantations in developing countries, have similar mean annual increments (MAI) in volume of 10–20 m^3 ha^{-1} $year^{-1}$ (Table 6.2; FAO 2000b). However, many species can achieve much faster growth rates. For example, *Eucalyptus grandis* can yield 40–50 m^3 ha^{-1} $year^{-1}$ and under good conditions with advanced tree improvement it can yield much more. Other widely planted tropical hardwoods including *Casuarina equisetifolia*, *Tectona grandis*, and *Dalbergia sissoo* have MAIs of less than 15 m^3 ha^{-1} $year^{-1}$ and frequently under 10 m^3 ha^{-1} $year^{-1}$ (FAO 2001a).

Plantation yield, whether it is measured in biomass or volume, is a function of the species' genetics, the silvicultural practices used, and site factors. Climate and site have a very large impact on growth rates. The humid tropics and more fertile sites are more conducive to higher growth rates than locations with long dry seasons or infertile or degraded soil. On many sites in India, for example, teak frequently has an MAI of only 4–8 m^3 ha^{-1} $year^{-1}$, partly because of drought combined with poor soils (FAO 2000b). In Costa Rica, in contrast, on good sites teak yields an average of 18 m^3 ha^{-1} $year^{-1}$ (Barquero 2004). Some species, such as *Gmelina arborea* and some of the *Eucalyptus* species, are very site sensitive. The average yield of *Gmelina arborea* worldwide is 25 m^3 ha^{-1} $year^{-1}$, while in Costa Rica it yields an average of 40 m^3 ha^{-1} $year^{-1}$ (Barquero 2004). *Pinus* species, in contrast, generally tolerate adverse conditions better.

Both tree breeding and silviculture have improved growth rates of several industrial species of eucalypts and pines. Good examples are *Eucalyptus grandis* and *E. urophylla* in Brazil (FAO 2000b). Much genetic improvement has been advanced by private companies, especially for the most frequently used species of pines and eucalypts. At Aracruz in the state of Espiritu Santo in southern Brazil, hybrids of *E. grandis* and *E. urophylla* (called "*urograndis*") yield an average of 44 m^3 ha^{-1} $year^{-1}$ and can be harvested at 7 years for pulp, or at 15 years for sawn wood (C.A. Roxo, pers. comm., www.aracruz.com.br).

Research on other species, including indigenous plantation trees, is ongoing at universities and other research institutions. For some native species in the humid tropics, genetic improvement has advanced to the stages of

Fig. 6.1. A progeny trial of *Vochysia guatemalensis*, at CATIE, Turrialba, Costa Rica. *Vochysia guatemalensis* is a native species broadly used in reforestation in the humid lowlands of Costa Rica. (Photo: F. Montagnini)

trials of seed origin and progenies, the first step in the domestication of indigenous species (Mesén et al. 2000). For example, for *Cordia alliodora*, *Vochysia guatemalensis*, and a few other native species in Central America, CATIE in Costa Rica has determined the best provenances (specific origin of the seed in a region or locality in a given country) that suit most planting conditions. In addition, progeny studies have helped to assess specific sources of seed (parent trees growing in a seed orchard or collection) for selected species such as *Vochysia guatemalensis*, *Acacia mangium*, *Eucalyptus grandis*, and other species of interest (Fig. 6.1).

Genetic diversity of a particular tree species is an important consideration in tropical plantations. Genetic diversity within a species helps to ensure that the species will be able to adapt to stresses and changing environmental conditions. The tendency among plant breeders is to select for individuals with a few desirable characteristics. Usually these characteristics are rapid growth and good form. Characteristics that are often neglected include: large roots; efficient association between roots and mycorrhizal fungi; production of root exudates that improve nutrient recycling; strategic reproductive patterns that take advantage of natural seed dispersal processes and avoid seed predation; high resistance to insects and disease; and allelopathic repression of other plants. When selecting individuals for reforestation using native or non-native species, it is important that the characteristics necessary for survival in the wild are included in the genome so that if it is not feasible to fertilize the plantations and spray them against insect attack, they will still be able to survive. When selecting trees to grow in specific sites such as in degraded lands, other characteristics that may be selected for should include the ability to grow on acid soils, withstand drought, and grow at low nutrient concentrations.

The length of the rotation period strongly affects both end-use and economics. Many fast-growing *Eucalyptus*, *Acacia*, *Casuarina* species, and *Gmelina arborea* are grown on short rotations of under 15 years, as they are used primarily for pulp or fuelwood. Usual rotations in Kenya for *E. grandis* are 6 years for domestic fuelwood, 7–8 years for telephone poles, and 10–12 years for industrial fuelwood (FAO 2000 b). In Brazil this species is largely grown for pulp or charcoal on 5- to 10-year rotations. Species being grown for high-value saw logs usually have longer rotations; teak (Tectona grandis) is grown in Asia on 50- to 70-year rotations (although in Costa Rica and other regions in Latin America much shorter rotations, about 25 years, are expected); and high-value conifers such as *Araucaria angustifolia* are grown on 40-year rotations. Generally, pines are grown on medium-length rotations of 20–30 years, unless grown solely for pulpwood, when shorter rotations may be adopted.

An important rule for deciding on the length of a rotation for plantations is the following: as long as diameter growth continues steadily, or does not decrease significantly, the yield will increase exponentially. The landowner's rate of return is increasing. A landowner with a stand of trees that average 50-cm diameter may be tempted to cut them and replant with new seedlings. The landowner should realize that by waiting, the yearly biomass increment (and hence profit) will increase much faster than the increment from newly planted seedlings. Once a forest stand has been cut and secondary succession begins, primary productivity drops way down. The reason is that biomass increases exponentially as diameter increases linearly. An example is as follows, assuming that tree trunks have a parabloic shape: A tree of 20 cm diameter and 10 m height that is growing at a rate of 2 cm diameter and 50 cm height year^{-1} (assuming a wood density of 1.0) will increase its weight by 42 kg year^{-1}. However, a tree of 40-cm diameter and 20-m height growing at the same rate will gain 164 kg year^{-1}.

Biomass increment of trees can be estimated by various means. If we assume that the trunk of the tree has a parabolic shape, then volume of the stem (PV) equals one-half the basal area times tree height (Whittaker and Marks 1975). Production for a given time period is the biomass at the final time minus the biomass at the time of the first measurement. Biomass of trees can also be estimated using allometric equations, where biomass is calculated as a function of relatively easily measured parameters such as diameter at breast height (DBH) and height of trees. Allometric equations are developed from field measurements of tree DBH, height, and biomass. Several allometric equations for estimating biomass of tropical trees have been developed, both for forests and plantation species (Brown and Iverson 1992; Montero and Kanninen 2002; Pérez and Kanninen 2002). The forms of the equations vary widely, including various polynomials and models incorporating DBH and height of trees, but the most common equation is the power function $B=aDBH^b$, where B is biomass. The power function is most often fitted by linear regression on log-transformed data (Dudley and Fownes 1992).

Biomass increment may not be the only consideration in deciding the length of a plantation rotation. There are also economic considerations. The increase in yield due to delaying a cut must be weighed against having the money now and having it in the future, at a value discounted to the present.

6.2.2 Sustainability of Forest Plantations

The issue of sustaining productivity over successive cycles of plantation establishment, growth, and harvesting has been discussed by several authors. This may be especially critical in tropical regions with fast-growing tropical tree plantations that incorporate considerable amounts of nutrients in their biomass over a relatively short period of time. Plantation forestry can cause large reductions in soil fertility through the excessive removal of living biomass, particularly if nutrients in tree crowns are lost through harvest or site preparation (Fölster and Khanna 1997; Wadsworth 1997). The harvest of forest products represents a nutrient cost to the site (Wang et al. 1991). This problem can be particularly serious when plantations are established on poor soils.

Management strategies to conserve site nutrients and enhance plantation sustainability may include preferential planting of tree species that do not place high nutrient demands on the site (Wang et al. 1991; Montagnini and Sancho 1994). Large differences may exist in nutrient use efficiency among tropical tree species. For example, in studies in Puerto Rico, Wang et al. (1991) found that *Casuarina* spp. was twice as efficient as *Leucaena* spp. for nitrogen (N), three to four times as efficient as *Albizia* and *Leucaena* for potassium (K), and about twice as efficient as all of the studied species for magnesium (Mg). A major practical issue in the next few decades is going to be the question of how much wood can be produced per unit of nutrient by a given plantation species under a regime of continual exploitation. To design viable tropical plantations, focusing on efficient use of nutrients at a stand level may be as important as considering production rates (Wang et al. 1991; Montagnini 2000). Nutrient removals in harvest are especially critical for the productivity of successive rotations of plantations at the same site. Knowledge of plantation species nutrient content in plant parts can give guidance to management considerations applicable at harvest (Box 6.1).

Box 6.1

Management of site nutrients in plantation forestry

Losses of nutrients during harvest may far exceed the rate of their replenishment by weathering of minerals in soils or by input via precipitation, especially when rotations are short (Fölster and Khanna 1997). However, the amounts of nutrients in various tree tissues (foliage, branches, stems) differ substantially. Results of studies of plantation biomass and nutrients in a humid tropical Costa Rican site showed that the relative tree tissue nutrient concentrations for all species were foliage >branches >stems (Stanley and Montagnini 1999; Montagnini 2000). In tropical plantations, the nutrients in the tree crowns generally account for a higher proportion of the total aboveground nutrients in the stand than do those in other compartments. Though branches and foliage summed together represented only 25–35% of total tree biomass, they generally represented about 50% of total tree nutrients (Fig. 6.2). To reduce the nutrient cost of harvests, site tree tissue biomass conservation should be prioritized as follows: (1) foliage, (2) branches, and (3) stems. Leaving branches and leaves on the site at the time of harvest, rather than harvesting the whole tree, would typically reduce the nutrient cost of log harvest by one-half. Additionally, the slash left on the ground would act as mulch, helping to improve soil conditions, and potentially increasing the number of rotations before fertilization or fallow would be necessary. The amount of nutrients represented by branches or foliage that may be left behind varies between nutrients, species, and sites. Adjusting harvest regimes in consideration of the differing nutrient contents of these tissues can be an effective means of managing site nutrients (Wang et al. 1991; Montagnini and Sancho 1994; Fölster and Khanna 1997; Nykvist 1997; Stanley and Montagnini 1999; Montagnini 2000). Additional considerations for nutrient conservation would include, in the case of N, avoiding burning since N can be lost by volatilization before being incorporated into the soil.

6.2.3 Plantations of Native Tree Species

The main fast-growing, short-rotation species used in plantations are in the genera *Eucalyptus* and *Acacia*, and, to a lesser extent, *Gmelina*. Pines and other coniferous species are the main medium-rotation utility species, primarily in the temperate and boreal zones. All these species are generally exotics in the locations where they are planted. Use of native species is minimal. Exotic tree species predominate in both industrial and rural development plantations worldwide (Evans 1999). In some regions of the humid tropics, plantations using in-

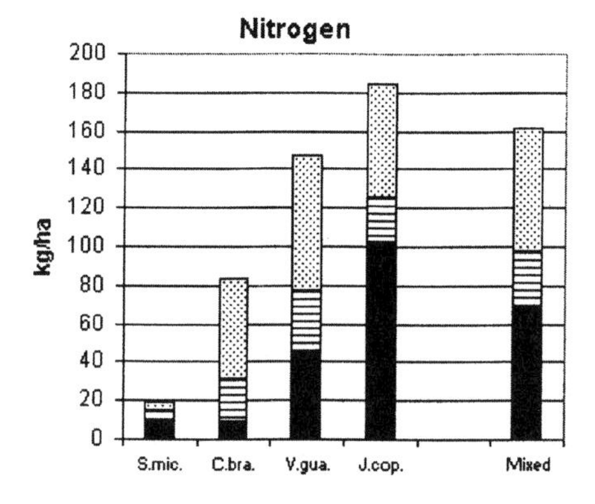
Nitrogen
kg/ha
200
180
160
140
120
100
80
60
40
20
0
S.mic.
C.bra.
V.gua.
J.cop.
Mixed

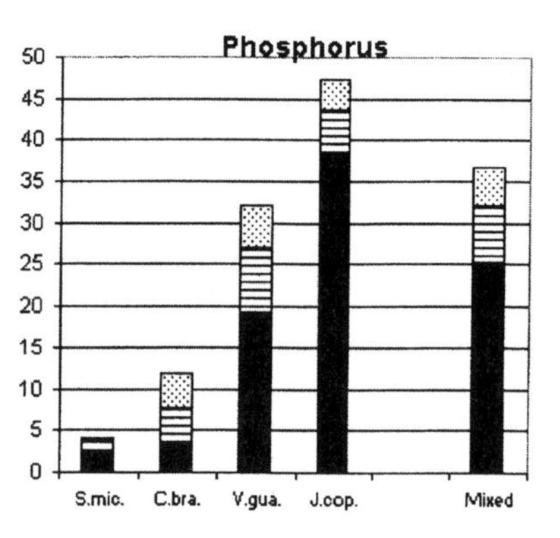
Phosphorus
50
45
40
35
30
25
20
15
10
5
0
S.mic.
C.bra.
V.gua.
J.cop.
Mixed

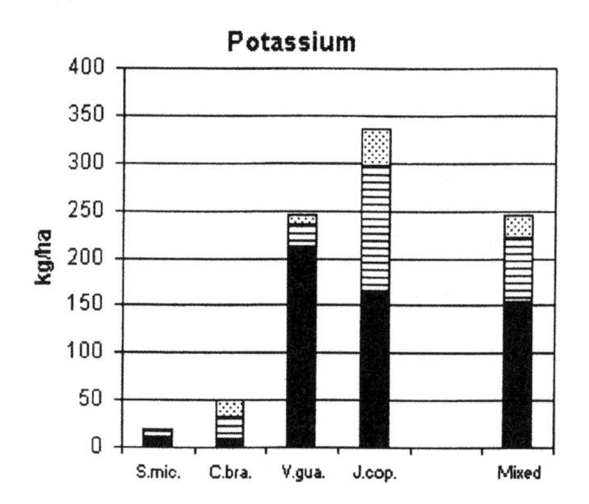
Potassium
kg/ha
400
350
300
250
200
150
100
50
0
S.mic.
C.bra.
V.gua.
J.cop.
Mixed

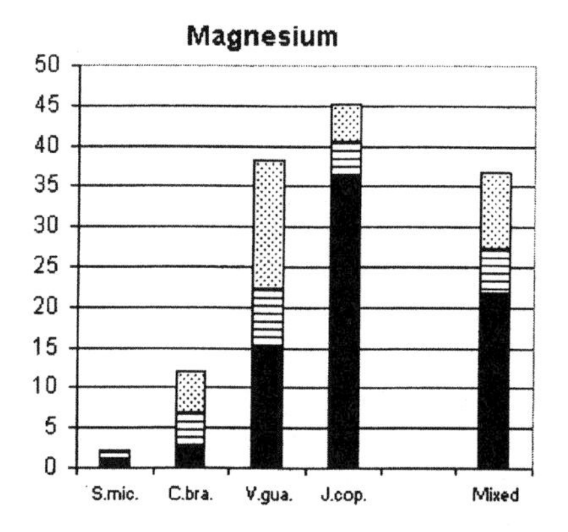
Magnesium
50
45
40
35
30
25
20
15
10
5
0
S.mic.
C.bra.
V.gua.
J.cop.
Mixed

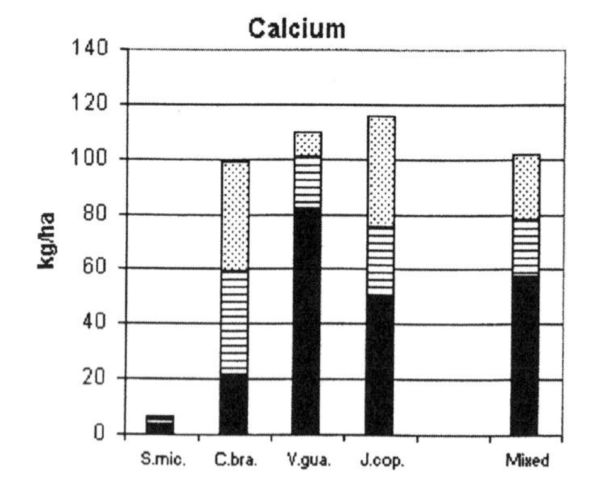
Calcium
kg/ha
140
120
100
80
60
40
20
0
S.mic.
C.bra.
V.gua.
J.cop.
Mixed

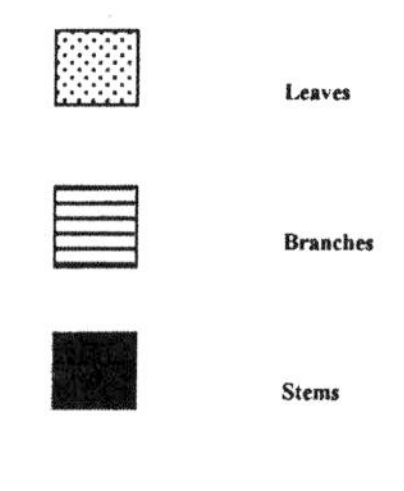
Leaves
Branches
Stems

digenous species are restricted for the most part to small and medium-sized farms practicing reforestation of degraded land in portions of their properties (Piotto et al. 2003a; Fig. 6.3).

Native trees can be more appropriate than exotics because (1) they are better adapted to local environmental conditions, (2) seeds and other propagules are locally available, and (3) farmers are familiar with them and their uses. Besides, the use of indigenous trees in productive systems helps preserve genetic diversity and serves as habitat for the local fauna. Disadvantages of the use of native species can be (1) uncertainty of growth rates and adaptability to soil conditions other than those where naturally found; (2) general lack of guidelines for management; (3) large variability in performance and lack of genetic improvement for most species; (4) seeds of native tree species are not as often commercially available and have to be collected; (5) high incidence of pests and diseases (e.g., the attack of *Hypsipyla grandella* on species of mahoganies and cedars, as mentioned earlier); and (6) lack of established markets for many species.

As a consequence of the disadvantages of native species, exotic trees are often and more generally preferred for reforestation because (1) there is generally more silvicultural information; (2) often it is possible to obtain seed of known genetic makeup and certified origin, and (3) there are generally well-established markets for plantation products. In several known cases the exotic species, not having local enemies, initially grow free from pests. However, in some cases a local pest finds the species and causes serious problems, as has been the case with the yellowing of china berry trees (*Melia azederach*) and stem cankers in Australian cedar (*Toona australis*), both in northeast Argentina and Paraguay; and stem rottening of *Acacia mangium* in Costa Rica (Montagnini, pers. observ.).

One of the strongest arguments for the use of native tree species in plantations is the high value of the wood and its increasing scarcity in commercial forests. Many native tree species of valuable timber grow well in open plantations, with rates of growth comparable or superior to those of exotic species in the same sites. Such is the case of *Vochysia guatemalensis* in Costa Rica, whose growth rate and value are similar to those of *Gmelina arborea*, and of *Terminalia amazonia* in Costa Rica and Panama, with growth rates and value similar to those of teak (Piotto et al. 2003a; Fig. 6.4).

Native species are often preferred for land restoration purposes, especially when the environmental services of plantations take precedence over their

◄

Fig. 6.2. Total aboveground nutrient content per hectare for *Jacaranda copaia*, *Vochysia guatemalensis*, *Calophyllum brasiliense*, and *Stryphnodendron microstachyum* grown in pure plantation, and a mixture of the four species at La Selva Biological Station, Costa Rica. Plantations were 3–4 years old at the time of nutrient and biomass determinations (Montagnini 2000, reprinted with permission from Elsevier Publishers)

Fig. 6.3. *Virola koschnyii* is a native species that is preferred by farmers for reforestation in the humid lowlands of Costa Rica due to its fast growth, good form, and good market value. This plantation is 10 years old. (Photo: F. Montagnini)

productive value. Box 6.2 describes PRORENA, a large reforestation project with native species in Panama that is developing knowledge of the potential of several native species for land restoration in the Panama Canal watershed and other regions of Panama.

Box 6.2

The Native Species Reforestation Project (PRORENA) (PRORENA 2003)

Since January 2001, the Center for Tropical Forest Science (CTFS) at the Smithsonian Tropical Research Institute (STRI) and the Tropical Resources Institute at the Yale School of Forestry and Environmental Studies have jointly led the Native Species Reforestation Project (PRORENA), a project directed at developing viable strategies for restoring diverse, native forest cover across the tropics. Based at STRI's facilities in Panama City, PRORENA seeks to develop solutions first in the Republic of Panama, and then to extend its impact throughout the Neotropics, and across the worldwide network of CTFS plots. PRORENA's efforts are guided by the vision of establishing diverse native forest cover across extensive areas of deforested tropical lands, and demonstrating that large-scale ecological restoration in the tropics is technically feasible, socially attractive, and financially viable.

Like most tropical nations, Panama is home to a bewildering diversity of tree species. For most species, even the most basic information necessary for their reproduction and management is lacking. A resulting paradox is that the most valuable native timber species are among those under the greatest threat of local extinction. Seed supplies are unavailable for most tree species. Since 2001, PRORENA has worked to create the information necessary to allow for the design of effective reforestation strategies, and to increase the local infrastructure and human capacities necessary to apply the results of this work.

PRORENA has initiated experimentation and field trials on more than 300 ha at nearly a dozen sites across Panama. These trials include more than 40 species, several of which appear in the literature simply with botanical descriptions and have never before been studied in a management context. However, PRORENA's most significant accomplishment has been the formation of a network of more than a dozen partner organizations dedicated to native species reforestation. These organizations worked in relative isolation before joining the Reforesters' Network, learning largely from their own experiences. PRORENA now captures the information generated by the efforts of each partner organization and makes this information widely available through seminars, reports, conference presentations, and the project website, improving the success and viability of reforestation efforts.

Systematic testing of species performance is conducted in trials replicated with a uniform design across a network of Core Research Sites covering a precipitation gradient of 1,000–2,700 mm year^{-1} and including a variety of soil conditions. Experimental selection trials include small plots of a total of 75 species from 38 geographically distinct populations. The selection trials involve planting a large number of species in order to select some that are most promising for each region. Within 1–3 years these trials will allow the identification of those species and populations that have the highest initial survival and growth rate both within and across sites and to consequently initiate more intensive testing of the more promising species. In addition to providing information on initial growth and survivorship, canopy morphology, and potential for biodiversity restoration for 75 poorly known species, the selection trials are located in order to maximize conservation and community impact. At two sites the trials are located on major thoroughfares, to serve as forestry demonstration plots for surrounding communities, at one site the trials are located within a deforested expanse of a National Park, and at all sites they are adjacent to existing forest fragments.

Fig. 6.4. A 9-year-old tree of *Vochysia guatemalensis* in an experimental plantation at La Selva Biological Station, Costa Rica. This is the most frequently planted native species for reforestation in the region, due to its good growth and economic value. (Photo: F. Montagnini)

6.2.4 Mixed Species Plantations

Species diversity can contribute to protection against pests and diseases in plantations such as those of cedar and mahogany (see Chap. 2). Species diversity can also improve resource use efficiency through complementarity. Most reforestation in the tropics is with monocultures of species such as pine, eucalyptus, and teak (Evans 1992, 1999). In monoculture plantations, the main interaction between individual trees is competition for nutrients, water, and light. In mixed-species plantations, differences in utilization of resources in space and time may benefit the ecosystem through greater primary production. When mixtures combine tree species that differ in growth requirements and production, they can reduce interspecific competition and can outyield monospecific stands (Kelty 1992). Stratified mixtures that combine rapidly growing overstory species with slow-starting but higher-producing species are likely to exhibit greater total productivity than pure stands of shade-intolerant species (Smith 1986). Mixed stands can improve the survival and growth of a species in nutrient-poor soils (Matthews 1989; Binkley et al. 1992).

Mixed species plantations have been established at several locations with varying results (Wormald 1992). However, results from a number of field experiments suggest that mixed designs can be more productive than monospecific systems (Burkhart and Tham 1992; Wormald 1992; Montagnini et al. 1995a; Montagnini and Porras 1998). In addition, mixed plantations yield more diverse forest products than monospecific stands, thereby helping to diminish farmers' risks in unstable markets. Farmers may prefer mixed plantations in order to diversify their investment and as a potential safeguard against pests and diseases, in spite of the technical difficulties of establishing and managing

Fig. 6.5. A 5-year-old mixed plantation at La Selva Biological Station, Costa Rica. The species are: *Jacaranda copaia* (the tallest trees), *Vochysia guatemalensis* (the second largest trees with round canopies), *Calophyllum brasiliense* (smaller tree in the lower layers), and *Stryphnodendron microstachyum* (not shown in the picture). (Photo: F. Montagnini)

Table 6.3. Biomass and carbon sequestration by 12 tree species in pure plots and in mixtures of four species each and their respective estimated rotation length at La Selva Biological Station, Costa Rica. Plantations were 6–6.5 years old at time of thinning for biomass determinations. Biomass values are means of four replicate plots per treatment (standard errors between parentheses). (Adapted from Shepherd and Montagnini 2001)

	Aboveground biomass		Mean annual stem increment (Mg/ha/yr)	Mean annual carbon sequestration (Mg/ha/yr)	Estimated rotation length (years)
	Stems	Total			
Plantation 1					
Calophyllum brasiliense	20.9 (0.9)	43.3 (1.6)	3.5	1.7	25
Jacaranda copaia	82.3 (7.5)	83.8 (7.7)	13.7	6.9	12
Vochysia guatemalensis	80.5 (4.1)	91.2 (9.8)	13.4	6.7	15
Four-species mixture	78.1 (7.4)	90.1 (8.1)	13.0	6.5	18
Plantation 2					
Albizia guachapele	21.4 (4.1)	28.2 (5.7)	3.8	1.9	20
Dipteryx panamensis	39.1 (4.2)	54.8 (4.5)	6.9	3.5	20
Terminalia amazonia	37.8 (3.3)	51.2 (4.2)	6.7	3.3	20
Virola koschnyi	34.1 (2.6)	49.6 (2.5)	6.0	3.0	15
Four-species mixture	39.8 (5.1)	57.0 (7.6)	7.0	3.5	18.7
Plantation 3					
Genipa americana	7.11 (1.0)	9.27 (1.3)	1.5	0.8	20
Hyeronima alchorneoides	31.5 (4.4)	44.0 (5.2)	6.8	3.4	20
Pithecellobium elegans	19.3 (1.5)	23.8 (1.9)	4.1	2.1	20
Vochysia ferruginea	26.7 (1.8)	42.0 (2.9)	5.7	2.9	15
Four-species mixture	22.7 (2.8)	36.0 (4.5)	4.9	2.4	18.7

mixed plantations (Piotto et al. 2003b). Mixed stands may also support a greater variety of wildlife and contribute to higher landscape diversity.

Comparisons of productivity of mixed and pure species plantations have been made experimentally at several locations in the humid tropics (see Boxes 6.3 and 6.4).

Box 6.3

Studies on experimental mixed plantations at La Selva Biological Station, Costa Rica

A number of studies have been conducted in experimental plantations of native species in mixed and pure-species designs at La Selva Biological Station in the Caribbean lowlands of Costa Rica (Fig. 6.5). At 2–4 years of age, mixed plantations had greater growth and lower pest damage than pure stands for three of the twelve species tested, and there was neither damage nor differences between pure and mixed conditions for the other species. The costs of plantation establishment were lower for the slower-growing species in mixed than in pure stands (Montagnini et al. 1995a). In comparison with pure stands of the fastest-growing species, mixtures had relatively high yields, with values similar or even higher than pure stands. When plantations were 9–10 years of age, most species had better growth in mixed than in pure-species plantations. However, the slower-growing species grew better in pure than in mixed stands, both on an individual tree and on a stand basis. Mixed plantations (combinations of three to four species) ranked among the most productive in terms of volume (Piotto et al. 2003b). Studies on soil fertility carried out in the same experimental setting showed that mixed plantations had a more balanced nutrient stock in the soil: 4 years after planting, decreases in soil nutrients were apparent in pure plots of some of the fastest-growing species, while beneficial effects such as increases in soil organic matter and cations were noted under other species. The mixed plots showed intermediate values for the nutrients examined, and sometimes improved soil conditions such as higher organic matter (Montagnini and Porras 1998). The role of pure and mixed plantations on biomass accumulation and carbon sequestration was also examined, measuring biomass of trees that were cut when performing thinnings at 3 and 6 years of age. The mixtures of four species ranked high in comparison with the pure-species plots of faster-growing species (Table 6.3). The mixtures of four species gave higher biomass per hectare than that obtained by the sum of 0.25 ha of each species in pure plots (Montagnini and Porras 1998; Shepherd and Montagnini 2001).

Box 6.4

Mixed plantations of native and exotic species in the Brazilian Amazon

To compare the growth of monocultures with plantations of mixed species in the Amazon region of Brazil, Batmanian (1990) established replicated plots of *Eucalyptus torelliana*, *Acacia mangium*, and a mixture of six native species. For the first few years, the eucalyptus and acacia grew much faster than the native species. This may have been because the seeds were from parents selected to respond well to plantation conditions – direct sunlight and fertilization. Only one native species, *Schizolobium amazonicum*, an early successional legume, grew well during the first few years. After about 4 years, however, the acacia stands were infected with a parasite from the misletoe family that severely damaged or killed all the trees within 1 year. The eucalyptus stands stagnated because they were not thinned. In contrast, as the stands of mixed native species began to get larger, their rate of growth increased. The increase was due partly to decreased competition from herbaceous weeds and grasses, and partly to an increasingly efficient use of light and nutrients. A similar result was found in a comparison of pine and native species in Puerto Rico (Jordan and Farnworth 1982). Initially the growth of pine was greater, but, eventually, the native species were more productive.

As seen from these examples, mixed plantations can have many productive and environmental advantages over conventional monocultures. However, their main disadvantage lies in their more complicated design and management. Thus mixed plantations are often restricted to relatively small areas or to situations where diversifying production is a great advantage, such as for small farmers of limited resources (Box 6.5).

Box 6.5

Reforestation with mixed and pure plantations in the Rio Carazo watershed, Nicaragua (Piotto et al. 2004 a)

Recently, Nicaragua has begun to expand its reforestation programs in response to the increasingly rapid deterioration of the country's forest resources. In 1997 the Programa Socioambiental y de Desarrollo Forestal (the Social Environment and Forest Development Program, POSAF), financed by the Banco Interamericano de Desarrollo (International Development Bank, BID), began to promote reforestation and the development of sustainable production systems in the farms of the Rio Grande Basin of Carazo (in southwestern Nicaragua) and in other regions of Nicaragua. POSAF supplied technical and financial assistance for the development of forestry projects and operations across various collaborative institutions. With the species recommended by the National Forest Service of Nicaragua this program began to establish forestry plantations on the farms of the producers that were benefiting from the project.

By the year 2000, about 12,000 ha had been planted throughout the country. Plantations were established between the years 1997 and 1998. The average size of the plantations was 1.8 ha per property. There was a mixture of different production types; some blocks were composed of pure timber species or pure fuelwood species and some blocks were composed of mixtures of timber and fuelwood species in various arrangements. Farmers used more native than exotic species for reforestation. The species preferred by farmers for their growth were *Azadirachta indica*, *Caesalpinia eriostachys*, *Eucalyptus* spp., *Gliricidia sepium*, *Leucaena leucocephala*, and *Tectona grandis*. Farmers were willing to continue reforesting as long as they continued to receive government incentives. The implementation of incentive programs for reforestation was a key factor in encouraging the participation of small and medium farmers. The most common plantation system used by farmers was mixtures of timber and fuelwood species, both of which were planted and managed to satisfy domestic needs.

Some private companies are also willing to establish mixed plantation systems in a restricted portion of their property while dedicating most of their land to planting other species of higher commercial value. For example, on the peninsula of Nicoya, in the northern Pacific region of Costa Rica, trials have been established with 13 native species in mixed and pure designs at the property owned by Precious Woods of Costa Rica, a private company that has reforested the majority of their lands with pure plantations of *Tectona grandis* (exotic) and *Bombacopsis quinata* (native) (Piotto et al. 2004b). The land occupied by the study sites was previously used for cattle ranching. Measurements of diameter at 1.3 m and height done when the trees were 68 months old showed that native species grew better in the mixed plots. In Panama, a company called Futuro Forestal is buying abandoned cattle pastureland to establish pure and mixed plantations for foreign investors (Box 6.6).

Box 6.6

Futuro Forestal S.A.: balancing nature and business (www.futuroforestal.com)
Futuro Forestal is located in the Chiriqui region, Panama, on the Pacific side of the isthmus. Since its foundation in 1994, Futuro Forestal has engaged in reforestation with mostly native species that produce fine tropical hardwoods of high value, carbon credits, and non-timber forest products, providing investment opportunities to its clients. The company has been the first to sell carbon credits from reforestation as a business in Panama. They have established mixed plantations of six tropical hardwood species in parcels of different sizes. The selection of species and their spatial distribution is defined by a detailed site examination. An additional 35 species are planted for biodiversity and research reasons. *Terminalia amazonia* is the most promising species so far.

In the first 5 years after planting, there is almost constant supervision and manual weeding. After a year the trees are pruned at the beginning and the end of every rainy season. Two applications of organic fertilizer per year are given to the seedlings until they grow with satisfactory vigor. From the fifth year on, when the canopy of the trees starts closing, the intensity of the management is reduced.

Futuro Forestal reconverts abandoned cattle land to natural habitat, creating a diverse and dynamic, but essentially artificial tropical forest. The company also buys land already supporting early secondary forest. In this case the company conserves the existing vegetation and enriches parts of it to increase its commercial value. Other areas are maintained as biological corridors. At present, of the total managed area of 372 ha about 82 ha is protected secondary forest. The forest management has been certified by the Forest Stewardship Council (FSC).

Mixed plantations have also been used in forest restoration projects. For example, in Sri Lanka, on abandoned swidden adjacent to the Sinharaja Man and the Biosphere reserve, mixed plantations of dipterocarps and other species were established using the canopy of an 18-year-old *Pinus caribaea* (Caribbean pine) plantation as 'nurse' for the shade-tolerant rain-forest species. In this experimental setting it was found that the best environment for seedling establishment and growth for all species were the centers of the canopy strips (6–12 m). In other experiments using only native rain-forest species, seedlings grew better when planted within openings created by the removal of three rows of pine canopy. Using already established pine species as nurses could thus be a solution to the dispersal, weed competition, and pathogen/insect problems that rain-forest tree species encounter during initial establishment on sites previously cleared of forest (Ashton et al. 1997a, 1998).

In southwestern Costa Rica, mixed stands involving 41 native species were established in a private forest reserve of 145 ha of abandoned pastureland. In evaluations carried out when plantations were 5 years old, sun-loving species had grown at a rate of 3 m year^{-1} in height and were larger than 10 cm DBH (Leopold et al. 2001). Likewise, mixed species reforestation schemes have been practiced in Australia involving both native and exotic species (Kanowski et al. 2003), in the restoration of bauxite mines in the Amazon (Parrotta

Fig. 6.6. A 2.5-year-old mixed plantation of native species established by the Panama Canal Authority in the Panama Canal watershed in collaboration with the PRORENA project (notice person standing at front in the middle for scale). (Photo: F. Montagnini)

et al. 1997), and near the shores of the Panama Canal (PRORENA 2003; Fig. 6.6).

6.2.5 Plantations and the Conservation of Biodiversity

In spite of their recognized value as substitutes for wood from natural forests and their key role in restoring local ecosystem services, tropical plantations are often viewed in a negative light (Carnus et al. 2003). It has been claimed that monocultures of exotic plantations are no more diverse than monocultures of soybeans or other agricultural crops. Some even do not want to use the word "forest plantations", claiming that monospecific plantations are not truly "forests". While plantations in general support fewer native wildlife species than a native forest, they often support more diversity that other land uses in the same region (e.g., pastures, degraded land). Plantations also support a greater diversity of native plant species in their understories than agriculture or pasture systems.

It is becoming widely accepted that the conservation of biodiversity has to take place in managed landscapes as well as in protected areas (Pimentel et al. 1992; Brown and Lugo 1994; Guindon 1996; Lamb 1998; Harvey and Haber 1999). In many regions of the tropics the landscape consists of a complex mosaic of forest patches, pastures, and agricultural fields that are heavily influenced by human activity. Any efforts to conserve biodiversity within this managed landscape must be compatible with local livelihood needs and offer sustainable and economically attractive alternatives to local farmers. Strategies that provide various ecosystem services and fulfill local human needs, in addition to promoting the conservation of biodiversity, will have a higher chance of success (Cairns and Meganck 1994; Lamb 1998).

One strategy that can potentially facilitate the maintenance or recovery of biodiversity within agricultural landscapes is the establishment of native forest plantations on degraded agricultural lands. In addition to providing a variety of economic and environmental services (such as timber production, carbon accumulation, soil protection, and land reclamation), plantations may help local biodiversity by facilitating regeneration of native tree species and providing habitats for forest animals (Parrotta 1992; Lamb 1998; Montagnini and Porras 1998).

A variety of management strategies can be used to increase diversity in plantation ecosystems, even those including exotic species. These strategies include: diversifying the number of tree species planted, leaving forest remnants in the landscape, and decreasing the intensity of management operations (Carnus et al. 2003). Management strategies that fall within the guidelines needed for forest certification help to ensure that plantation forests as well as native forests are managed in a way that promotes wildlife habitat.

6.2.6
Plantations in the Landscape

Other environmental concerns apply more to large-scale industrial monospecific plantations of exotic species. Concerns include their potential deleterious effects on soils (e.g., increases in soil acidity under pines) and decreases in water yield downstream (Cossalter and Pye-Smith 2003). Some plantation species, such as eucalyptus, are highly susceptible to fire because of the fuel that they can accumulate as litter on the ground under their canopies. A classic example from South Africa shows how dramatic declines in water yield actually occurred following planting of riverside areas with eucalypts, black wattle, and pines in the 1950s and 1960s, leading to the elimination of the plantations in the 1990s (Cossalter and Pye-Smith 2003). However, in other cases, eucalypt plantations have been shown to have beneficial effects on the ecosystem: for example, in the People's Republic of Congo, plantations of fast-growing eucalypts established on degraded savanna soils have been shown to have improved the soils by building up organic matter, and even encouraged the return of natural vegetation and wildlife to the site (Loumeto and Huttel 1997). In all cases, plantations have to be put into context with the rest of the landscape; the type of land use plantations are replacing (natural forests, pasture, agriculture); their main purpose (industrial, recovery of degraded land, etc.); and other benefits plantations provide (carbon sequestration, economic growth, and social functions such as greater income and employment) (Cairns and Meganck 1994; Evans 1999; Keenan et al. 1999; Sedjo 1999).

6.2.7
Plantations as a Tool for Economic Development

The economic and social benefits of plantations have been as much a source of debate as their environmental consequences. For example, it may be claimed that plantations generate employment, but it may also be argued that this is so only in the first phases of plantation establishment. Plantations can bring economic development to a country: exports may contribute to the balance of payments, taxes may flow to the national treasury, and plantations may generate jobs and prosperity. The governments of several countries – most notably China, Japan, and South Korea – have invested in medium- and long-rotation plantations precisely because they see them as a means of creating jobs and stimulating rural development (Cossalter and Pye-Smith 2003).

For example, in Brazil, Aracruz Celulose is the world's leading producer of bleached pulp. The company is responsible for 30% of the global supply of the product, used to manufacture high-value-added goods such as tissue, and printing, writing, and specialty papers. Aracruz's forestry operations in the states of Espírito Santo, Bahia, Minas Gerais, and Rio Grande do Sul involve some 242,000 ha of eucalyptus plantations, intermingled with 121,000 ha of company-owned native forest reserves. The company exports almost all of its production, currently 2.4 million tons/year, and is one of the largest earners of foreign exchange in the Brazilian manufacturing sector, making a substantial contribution to the country's balance of trade and overall development (www.aracruz.com.br). On the other hand, this type of large-scale plantation development has attracted much criticism from an environmental point of view. Concerns about the potential deleterious effects of eucalypts on water yield downstream have been expressed by many authors, as discussed in the previous section. In addition, Aracruz's hybrid eucalypt plantations occupy land that used to harbor Atlantic rainforest, one of the most endangered forest ecosystems of the world. However, deforestation was very severe in that portion of the Atlantic forest region, and plantations were set on previously deforested, degraded pastureland. The company's environmental division also emphasizes its use of native forest as buffers interspersed in its plantations as a way of mitigating the negative environmental consequences of growing large monoculture plantations of exotic trees.

From a regional development point of view, plantations may also result in economic losses. Some industrial plantations are established with financial support from the state, and therefore public taxes are used to finance private economic ventures. Economic benefits of plantations must be felt publicly so that these investments of public monies are justified. The advantages of such economic development must also outweigh any potentially negative environmental effects for plantations to be considered a successful development venue. In Box 6.7, we present a case study where plantation forestry brought economic development to a region, reconciling conservation objectives with economic development.

Box 6.7

Plantation forestry in Misiones, Argentina: productivity and conservation (Fernández et al. 1997; Montagnini et al. 1997, 1998, 2001; Eibl et al. 2000; Gobierno de la Provincia de Misiones, Argentina 2003 a, b)

The province of Misiones in northeastern Argentina (25–28 °S, 53–56 °W, 100–800 m elevation), has an area of approximately 30,000 km^2, less than 1% of the country total; however, it harbors almost 40% of the biodiversity and produces over 70% of the country's timber. The subtropical forest of Misiones formerly covered more than 100,000 km^2 in regions of Argentina, Paraguay, and Brazil, but has been reduced to less than 10% of its original size. The Paranaense forest is one of the most diverse ecosystems of both Argentina and Paraguay. In its mature form, the Misiones forest contains an average of about 100 tree species per hectare. The species composition varies with geographical location: for example, the presence of *Araucaria angustifolia* (Bert.) O.K. (pino Paraná) in association with tree ferns is restricted in the northeast part of the province (at higher elevations), while *Aspidosperma polyneuron* Müll. Arg. (palo rosa) in association with palmito (*Euterpe edulis*) only occurs in the north. The complex forest structure includes trees attaining heights of up to 40 m, with no single species reaching importance values (average of relative abundance and relative frequency) of more than 8%, as well as a dense understory of lianas, tree ferns, palms and bamboos, small shrubs, and herbs.

The extent of subtropical forest in Misiones is currently about 1,800,000 ha, not including plantations. About 200 local producers own forest subjected to National Forest Management Plans, for a total area of forest under management of almost 500,000 ha. Timber extraction is carried out using diameter limits established for each individual species (see Chap. 5).

About one-third of the total area of forest is in national and state parks. There are a total of 53 Protected Areas of different categories, including national parks and reserves, 17 provincial parks, 16 private parks and wildlife refuges, natural monuments, two world heritage sites, one biosphere reserve, and one area of conservation and sustainable development: the Green Biological Corridor (Corredor Verde). In general, in these protected areas all the ecosystems of the Paranaense forest have been preserved.

A forest inventory completed in 1850 revealed by then a total area of natural forests of 2,600,000 ha, but as a result of forest exploitation and clearing for establishing plantations and agriculture, by 1977 this size was reduced to 1,222,000 ha. This was largely a result of government incentives for commercial plantations of pulpwood (principally *Pinus elliotii* and *Pinus taeda*), and cash crops such as soybeans, yerba mate (*Ilex paraguariensis*), and tea. The first tree plantations were established in the 1920s with the native *Araucaria angustifolia*. Pines were first planted in 1948, and today they dominate the Misiones landscape, mostly to the east of the province on the coast of the Paraná River. The first cellulose and paper plant was installed in the 1950s near the town of Eldorado on the Paraná River. The cellulose and paper industry soon consumed the natural and planted supplies of araucarias, and thus the plantations of pines and eucalypts were started to supply the fiber demand. By the mid-1980s, there were already over 150,000 ha of forest plantations, and as a consequence the forest cover of the province, including plantations, was increased to 67% by 1985. [The coverage was somewhat less (54%) in 2003.]

Due to the relatively low prices paid by the cellulose industries for pulp wood from the plantations, saw mills that cut timber for boards proliferated especially in the northern part of the province and spread later to the rest of the province. Advantages for the development of the pulp and timber industry in Misiones were the high yields and relatively short rotation times of plantation species; availability of high-quality labor; and natural forest resources of high diversity. In recent years, apart from its tradition as a province that supplies timber products for the rest of the country and for exports, Misiones has incorporated the concept of "multiple use" of its forest resources, including the use of non-timber forest products, as well as taking advantage of environmental services and recreational uses of forests.

Today, Misiones contains about 330,000 ha of planted forest. In 2003, Misiones' total production of roundwood from plantations was about 5,000,000 m^3, representing about 70% of total production, the rest coming from natural forests. Most plantations consist of exotic species such as *Pinus* spp. (*elliotii*, *taeda*, and others), *Melia azederach* L. var. *gigantea* (paraiso), *Eucalyptus* spp., *Paulownia* spp., *Toona ciliata*, *Grevillea robusta*, and the native *Araucaria angustifolia* (Bert.) O.K.

The area cultivated with native tree species (*Araucaria angustifolia* and a few others) constitutes less than 10% of the total. This situation is partially due to insufficient information regarding adequate silvicultural methods for the establishment and management of native species. Yet, alternatives for land use are clearly necessary to support current economic and ecological needs: inappropriate land use and management with the proliferation of shifting agriculture or the use of inadequate soil management strategies in plantation management has often led to soil degradation and to subsequent abandonment of lands in the region.

In 1989, the Yale School of Forestry and Environmental Studies (FES) signed an agreement with the School of Forestry, National University of Misiones (UNAM), to carry out collaborative studies in reforestation and agroforestry. Experimental pure and mixed plantations with native timber species on degraded land and enrichment experiments in degraded forests were established using native trees of economic value. Additionally, understanding of the use of native tree species in agroforestry combinations with commercial crops and their use in enrichment planting of degraded and secondary forest increased (see next sections in this chapter).

In addition, UNAM has recently started planting native tree species on degraded land in the Arroyo Pomar watershed, a tributary of a lake in the municipality of Eldorado. Local people with limited economic resources help care for and manage these plantations, which are expected to provide forest products and improve the environmental conditions (soil, water quality) of the watershed. The species used were chosen based on results from the Yale FES/UNAM research. The Pomar River restoration project will be extended by collaborating with other municipalities of the province of Misiones, to plant native species in degraded watersheds as part of communal projects to provide tree products and improve local environmental conditions.

Currently the difficult economic and social situation in Argentina (in spite of recent economic development) makes urgent the need for finding practical solutions to increase productivity and decrease negative impacts on existing resources. Local farmers are willing to try new alternatives for ecosystem restoration, some of which may include agroforestry combinations with the most common cash crops, and forest enrichment practices in impoverished secondary forest. Agrosilvopastoral systems with improved cattle breeds are helping many farmers to accelerate returns on plantations, especially pines.

Forestry development in the province of Misiones has contributed to increasing productivity and diversifying of the economy of the province while respecting environmental constraints. For example, plantations for pulp and timber keep expanding; however, they are generally established on relatively flat land. Most of the upper elevations of watersheds and riparian areas in the province are still covered by natural forest (Fig. 6.7). The relatively large proportion of forest in protected status ensures conservation of the diverse ecosystems of the region. Their use in ecotourism also contributes to diversifying the economy of the province. The challenge is to ensure that forest management respects the management plans; to check illegal logging; and to control encroachment of people on the forest, especially from the poorly protected borders with Brazil and Paraguay.

Fig. 6.7. Pine plantations usually occupy lower elevations in Misiones, Argentina, while natural forests are retained in the upper elevations and along rivers. (Photo: F. Montagnini)

6.3 Agroforestry

Agroforestry is a new name for a set of old practices. There have been many definitions of agroforestry, but they can be summarized in the following: "Agroforestry is a land-use system in which woody species are grown intentionally in combination with agricultural crops or cattle on the same land, either simultaneously or in a sequence. The objective is to increase total productivity of plants and/or animals in a sustainable manner, especially under levels of low technical inputs and in marginal lands. It involves the social and ecological integration of trees and crops" (Nair 1989).

Inclusion of woody components in a production system can provide benefits from the tree products and functions (timber, fuelwood, leaf mulches, the fencing function in a living fence, etc.; Fig. 6.8) and from their potential ecological advantages, especially their nutrient cycling abilities. The choice of a tree species will often depend on whether both productive and ecological advantages can be achieved in the same system, and in some cases one prevailing function, either productive or environmental, may be desired.

Fig. 6.8. A living fence of *Gliricidia sepium* near Siquirres, in the Atlantic lowlands of Costa Rica. The prunings from this nitrogen-fixing tree can be used for mulching or for fodder. (Photo: F. Montagnini)

6.3.1
Most Frequently Used Agroforestry Systems

There are many types of agroforestry systems. Some are *sequential*, such as the taungya systems where trees are intercropped with annual species during the first years of plantation establishment (Fig. 6.9). After the plantation canopy closes, the crops are no longer maintained and the system is managed as a plantation. This serves to decrease costs of weeding of the plantation, and allows the farmer to use the land in the plantation for the crops (Jordan et al. 1992). An example of a taungya system in Thailand, developed primarily to stimulate teak reforestation, is presented in Box 6.8.

Other sequential agroforestry systems involve shifting cultivation, where trees are planted in the fallow fields to speed up the recovery of soil fertility and to obtain the tree products. Thus crops and trees share the same space only during a portion of the whole productive cycle. Traditionally, shifting cultivators have encouraged the presence of certain tree or herb species in fallows to restore site fertility, suppress weeds, and increase economic yields. Several types of traditional enriched fallows involve planting or tending selective species for fruit, fuelwood, or timber for local consumption and markets (Nair 1990; Kass et al. 1993; see also Chap. 5).

Fig. 6.9. A taungya system of rubber trees intercropped with peanuts in Dak Lak province, Vietnam. (Photo: Hoang Dinh)

Box 6.8

Taungya systems for reforestation with teak in Thailand (Gajaseni 1992; Gajaseni and Jordan 1992; Phothitai 1992; Preeyagrysorn 1992; Kaothien and Webster 2001)
In the 1950s, Thailand became increasingly involved in international trade. One of its greatest resources was the northern teak forests. In an effort to balance international debt payments, Thailand increased exploitation of its forests. Logging, both legal and illegal, eliminated huge areas of forest. Between 1961 and 1988, forest cover was reduced from 53 to 28% of the total area of the country.

To remedy the problem of an inadequate supply of teak and the need for employment of landless peasants, a modified version of taungya was introduced in Thailand. Taungya is a land management system with origins in Burma (Myanmar). The British Colonial foresters encouraged peasant farmers to reforest logged areas with teak, allowing them to cultivate crops for the first few years among the teak seedlings. The system was resurrected in 1975 when the Cabinet of Ministers directed the Royal Forestry Department to reallocate land in degraded national forest reserves through the Forest Village System.

If a peasant family joins a Forest Village System they are allotted 1.6 ha of land each year. The land is to be cleared and planted with forest trees. Cultivation of crops such as upland rice is permitted between the tree seedlings. An additional 0.16 ha is given for a dwelling and home gardening. Water, electricity, education, and health care are provided. Two laborers from each family are recruited to work in plantation activities. This employment adds substantially to family income.

The Forest Village Systems have only been partially successful. While large areas of degraded land have been reforested with valuable timber species, the system is not without problems. Nutrient losses occur during the cropping period, but they are small compared to the total stocks in the ecosystem. These losses probably do not affect the growth of trees. However, nutrient loss due to clear-cut harvesting at the end of each rotation of teak can deplete stocks to levels insufficient to ensure satisfactory production during subsequent rotations.

Social and economic problems also occur within some taungya systems. A frequent source of discontent amongst the peasants is that they are unable to own property. Some villagers view their participation in the system as only temporary until they can somehow own their own land. This causes conflict with the system managers who would prefer the villagers to remain permanently. A nomadic tradition in some indigenous groups is also an obstacle to integrating them into taungya village systems. Other problems result from competition between trees and crops and the peasants not having strong incentive to ensure success of the tree plantation. They may have more interest in ensuring good crop growth from which they will derive personal benefit. Sometimes the peasants deliberately damage the plantation trees in order to favor the growth of crop species. Another problem in plantations of long-lived species such as teak is the long time interval during which there is no economic income from the land on which the trees are planted. An effort to remedy the problem is being tried in several of the forest villages. Fruit trees are interplanted among the timber species. The fruit trees begin to produce shortly after the cultivation of rice ceases and the economic production can help bridge the gap until the time that the timber trees are harvested. Fruit of trees such as tamarind (*Tamarindus indica*) are popular in the markets of Bangkok.

Agroforestry systems can be *simultaneous* when trees and crops or animals are in combination for the whole duration of the productive cycle. Some examples are permanent crops such as shade-grown coffee and cacao (Figs. 6.10 and 6.11), home gardens, alley-cropping systems, and agrosilvopastoral systems (combinations of trees with animals and pastures or agricultural crops). In alley cropping, annual crops are grown between hedgerows of leguminous nitrogen-fixing shrubs and trees, which are periodically pruned to prevent shading of companion crops. The prunings can then be used as mulch and green manure to improve soil fertility and produce high-quality fodder. Alley cropping is regarded as an improved bush–fallow system with the following potential advantages: (1) cropping and fallow phases are combined; (2) cropping periods are longer, and land is used more intensively; (3) soil fertility is effectively maintained with the use of species selected for that purpose; and (4) the need for external inputs is reduced (Kang and Wilson 1987).

Several studies have shown the ecological and economic advantages of combining nitrogen-fixing trees and timber or fruit trees with perennial crops such as coffee and cacao. For example, in Costa Rica, Beer (1988) compared the annual nutrient return in litter fall and prunings in systems of coffee with a nitrogen-fixing leguminous tree, *Erythrina poeppigiana* (poró), and a timber tree of good commercial value, *Cordia alliodora* (laurel). Both trees are common in agroforestry systems with perennial crops in Latin America. The amounts of nutrients recycled by the associated trees reached the recommended levels of fertilizer required for coffee production. The system including laurel was preferred by the farmers because of its timber value. Several other examples of combinations of trees and perennial crops have been described (Box 6.9). Their advantages are generally both ecological (the contribution of trees to sustainability of the systems) and economic (product diversification, more efficient use of land).

Fig. 6.10. Coffee in combination with *Leucaena leucocephala*, and pepper (*Piper nigra*) growing on top of the *Leucaena* stems in Dak Lak province, Vietnam. The *Leucaena* provides shade and nitrogen fixation and the pepper is grown for additional cash income. (Photo: Hoang Dinh)

Fig. 6.11. Cacao growing under the shade of fully grown, already producing cashew nut trees in Dak Lak province, Vietnam. (Photo: Hoang Dinh)

Box 6.9

Agroforestry systems of a perennial crop (*Ilex paraguariensis*) and timber trees in Misiones, Argentina

In the province of Misiones, northeastern Argentina, agroforestry systems of timber trees and perennial cash crops are becoming increasingly common. For example, agroforestry designs with native trees of economic value and cash crops have been proposed as land use options for a heavily deforested area of the province (Eibl et al. 2000). In recent research, the productivity of *Ilex paraguariensis* (South American holly or yerba mate, Aquifoliaceae) in association with indigenous trees *Enterolobium contortisiliquum* (timbó, Leguminosae, an N-fixing tree), and two timber species, *Balfourodendron riedelianum* (guatambú, Rutaceae) and *Tabebuia heptaphylla* (lapacho negro, Bignoniaceae) was evaluated on two private farms. Five years after planting, the tree species were 3.5–8 m high and 3–8 cm in diameter at breast height, and the yerba mate produced its first harvest. Additionally, production from associated subsistence crops covered the annual needs of the farmer. Considering the short-term returns in terms of subsistence annual crops, the medium- to long-term profits from yerba mate harvests, and the long-term returns from timber, the systems proposed here appear to yield an attractive sequence of products. Apparently these systems have value at both the subsistence and the commercial production levels, thus making them an interesting alternative for the small farmers of the region.

Home gardens are agroforestry systems for the production of subsistence crops for the gardener and his/her family, with or without the addition of cash crops. They can be located immediately surrounding the home or slightly further away, but still near the residential area. It is claimed that home gardens originated in prehistoric times when hunters and gatherers dispersed seed of highly valued fruit trees close to their homes (Soemarwoto 1987; Boxes 6.10 and 6.11).

Box 6.10

Home gardens: an ancient agroforestry system

In the Near East, home gardens are documented in paintings dated 3,000 years b.c., and the practice continues in modern times. There are several well-documented examples of home gardens from Java, and they are also common in other parts of Indonesia, Malaysia, India, Africa, and Latin America. They can contain great species diversity with many life forms varying from climbers to tall trees and vines, creating a forest-like, multistory canopy structure. The canopies of most home gardens consist of two to five layers. Usually, there are no rows, blocks, or definite planting distances among components. Chemical fertilizers are generally not used, but rather dung, household wastes, and pruning residues are used instead. The use of species with anti-pest properties is also a widespread practice (Fernandes et al. 1989; Michon et al. 1989). Home gardens generally have stable yields and great variety of products, allowing continuous or repeated harvests during the year under a low-input system.

In western Java, the average size of a home garden is <0.1 ha, with an average of 19.0 and 24 species per garden in the dry and wet seasons, respectively. The size of home gardens first increases, then decreases with altitude. The highest number of species occurs at 500–1,000 m. Poor people tend to grow more staples, vegetables, and fruits, while wealthier people tend to grow more ornamentals and high-economic value cash crops. More subsistence crops are grown in remote areas; more cash crops are grown near cities. Culture and tradition influence composition: e.g., more medicinal plants are found in west Java, while tobacco and coffee are more commonly grown in Muslim districts of southern Ethiopia. Animals are found in most gardens, but pigs are not found in Muslim home gardens. In west Java with intense rains, fishponds are usually present (Soemarwoto 1987).

Box 6.11

Sustainability of home gardens

Home gardens can be sustainable production systems, but usually only under low-input and low-yield conditions. For example, the home gardens of the Chagga, on Mt. Kilimanjaro (Tanzania), represent ecologically sustainable land-use systems, but their productivity is relatively low and needs to be increased if they are expected to support larger populations (Fernandes et al. 1989). Migration of young people to urban areas has disrupted the traditional transmission of the knowledge and experience required for the successful management and perpetuation of the complex multi-cropping system. Availability of fertilizers has decreased the need for organic manures, thus greatly reducing labor inputs in home gardens and therefore reducing nutrient recycling processes. If home gardens are to be used to raise the standard of living of people to satisfactory levels, the question arises whether the yield and the income can be increased without sacrificing their sustainability.

Agrosilvopastoral systems – the combination of timber, fuelwood, or fruit trees with animals, with or without crops – are practiced at many scales. A large-scale system may include timber plantations with grazing to control weeds and to obtain a more immediate return from the sale of animal products. Cattle-raising can also complement subsistence agriculture, with animals often integrated in home gardens or in systems of fodder production to feed animals in stables (Montagnini 1992). In some regions, the incorporation of trees – especially multiple-purpose trees (MPTS) – can change cattle-raising from an inefficient use of land to a more ecologically and economically feasible activity. The incorporation of trees can improve system productivity either by increasing pasture yields through more efficient nutrient recycling or through the production of fodder from leaves and fruits (Gill et al. 1990; Cobbina 1994/1995). The introduction of cattle in plantations, generally when the trees are large enough so that cattle cannot damage them, can be a more efficient use of land than plantations alone. Cattle can also help to control weeds, as in a taungya system, and the early economic returns from cattle products help to pay for the costs of reforestation (Montagnini et al. 2003; Fig. 6.12). In some cases, young calves are preferred over adult cattle because they cannot damage the trees as easily. The smaller animals also decrease soil compaction problems. Some tree species are more preferred by cattle for browsing; cattle can also damage young trees when they scratch against them or strip their bark.

Fig. 6.12. Agrosilvopastoral system of a young plantation of a native timber tree (*Hieronyma alchorneoides*), grazed by young calves (beef cattle) on a farm near Horquetas in the Atlantic lowlands of Costa Rica. No pasture species were planted under the trees, the calves are grazing natural grasses. (Photo: M. Padovan)

6.3.2
Functions of Agroforestry Systems

Agroforestry systems are used in a variety of different ecological and economic conditions. For example, in fertile regions agroforestry systems may be very productive, yielding high quantities of crops such as cacao and rubber (Fig. 6.13). However, their greatest potential to increase productivity and sustainability of production systems is in degraded areas, in regions with soils of low fertility, or in semi-arid regions. Also, agroforestry systems have a potential to increase sustainability of agriculture and income to small farmers who lack adequate infrastructure or technical resources.

Agroforestry systems are especially important in regions where commercial fertilizers are expensive or unavailable, because of their ability to recover, recycle, or efficiently utilize nutrients. This ability is often linked to mechanisms associated with woody or perennial species that recycle nutrients mainly through litter fall and decomposition. While agroforestry systems can be profitable if established immediately after forest clearing, they often require a number of years to become profitable when established on degraded lands. For this reason, capital-limited farmers on poor soils may require subsidies to enable the establishment of agroforestry systems (Montagnini et al. 2000).

Fig. 6.13. Rubber trees intercropped with corn in a taungya system in Dak Lak province, Vietnam. (Photo: Hoang Dinh)

In contrast to regions with relatively fertile soils, where plantation forestry or intensive agriculture may be a preferred productive system, agroforestry may be a better alternative in regions with more nutrient-poor soils, or with poorer socio-economic conditions. For example, a taungya system may be a more acceptable system to farmers than plantation forestry where there is an immediate need to produce food or cash. Intensive agriculture such as coffee monoculture may be more economically profitable, but it requires high inputs which many small farmers cannot afford. However, coffee can be a productive alternative when mixed with shade trees which can contribute to nutrient recycling in a less intensive system with lower inputs.

Agroforestry systems also contribute to the conservation of biodiversity and carbon storage (Dixon 1995; Young 1997; Beer et al. 2003; Montagnini and Nair 2004). Agroforestry systems can play an important role in the conservation of biodiversity within deforested, fragmented landscapes by providing habitats and resources for plant and animal species, maintaining landscape connectivity (and thereby facilitating movement of animals, seeds and pollen), making the landscape less harsh for forest-dwelling species by reducing the frequency and intensity of fires, potentially decreasing edge effects on remaining forest fragments, and providing buffer zones to protected areas (Beer et al. 2003). Agroforestry systems cannot provide the same niches as the original forests and should never be promoted as a conservation tool at the expense of natural forest. However, they do offer an important tool for conservation and should be considered in landscape-wide conservation efforts to both protect and connect remaining forest fragments and promote the maintenance of on-farm tree cover in areas surrounding protected areas.

Agroforestry is important as a carbon sequestration strategy because of carbon storage potential in its multiple plant species and soil, as well as its applicability in agricultural lands and in reforestation. Proper design and management of agroforestry practices can make them effective carbon sinks. As in other land-use systems, the extent of C sequestered will depend on the amounts of C in standing biomass, recalcitrant C remaining in the soil, and C sequestered in wood products. Average carbon storage by agroforestry practices has been estimated to be 9, 21, and 50 Mg C ha^{-1} in semiarid, subhumid, and humid tropical regions (Schroeder 1994; Dixon 1995). Agroforestry can also have an indirect effect on C sequestration when it helps decrease pressure on natural forests, the largest sink of terrestrial C. Another indirect avenue of C sequestration is through the use of agroforestry technologies for soil conservation, which could enhance C storage in trees and soils (Young 1997). Agroforestry systems with perennial crops may be important carbon sinks, while intensively managed agroforestry systems with annual crops are more similar to conventional agriculture (Young 1997; Beer et al. 2003; Montagnini and Nair 2004).

6.4
Restoration of Degraded Tropical Forest Ecosystems

If plantation and agroforestry species are chosen with knowledge of their nutrient-use efficiencies and recycling capacities, they can be highly productive and be used in ecosystem restoration projects. The term restoration is generally used for those situations where the intent is to recreate an ecosystem as close as possible to that which originally existed at a particular site, while the term rehabilitation is more specifically used when enough structure is regained to recover at least some productive capacity. For example, the word restoration applies when the intent is to recreate a forest that will be as close as possible to the original forest that existed prior to a disturbance. In contrast, rehabilitation of degraded land often implies that the intention is to recover soils to productive capacity. An example would be to turn a degraded pasture into a commercial or subsistence-oriented agroforestry system or tree plantation.

According to a report by ITTO (2002), some 350 million ha of tropical forest land has been so severely damaged that forests will no longer grow back spontaneously, while a further 500 million ha has forest cover that is either degraded or has regrown after initial deforestation (Table 6.4). Such large areas of damaged forest are a cause of concern, but they also represent a potential resource of immense value. Degraded primary forests, secondary forests, and degraded forestlands usually exist in a complex mosaic that is constantly changing. Each of these conditions, however, has characteristics that must be taken into account when developing management strategies. In the following sections we describe strategies to restore or rehabilitate degraded primary or secondary forest and degraded forestlands.

Table 6.4. Estimated extent of degraded and secondary forests by category in tropical Asia, tropical America, and tropical Africa in the year 2000 (million hectares, rounded to the nearest 5 million). In tropical America, about 38 million ha is classified as secondary forests (second growth forests), while for the other regions it is not possible to distinguish between degraded primary forests and secondary forests. (Source: ITTO 2002)

	Asia (17 countries)	America (23 countries)	Africa (37 countries)	Total
Degraded primary forest and secondary forest	145	180	175	500
Degraded forest land	125	155	70	350
Total	270	355	245	850

6.4.1
Recovery of Degraded Forests

Common causes of disturbance leading to degraded primary and secondary forest are excessive wood exploitation; overharvesting of non-wood forest products; overgrazing; and destructive natural disturbances such as forest fires, storms, and hurricanes (ITTO 2002). As a result of such disturbances, forest structure is not significantly damaged, but there may be poor understory development and absence of young age classes of canopy species (especially of those that had been extracted from the forest). In some cases, there may be invasion by light-demanding species, such as grasses and bamboos. The regrowing forest may differ in species composition and in physiognomy from the original primary forest.

Degraded primary and secondary forests today represent a major source of forest products in several countries (see Chap. 5). For example, in Costa Rica, the area covered by these types of forest is estimated to be more than 600,000 ha. In the Philippines almost all the dipterocarp forests are now highly degraded. Similar situations are found in Sri Lanka, Nepal, the Brazilian Amazon, and West Africa. The World Bank has estimated that about 300 million people depend on degraded or secondary forests for their livelihoods (ITTO 2002).

Possible choices for the restoration or rehabilitation of degraded and secondary forest may include:

- Leave to regenerate (for example, as part of a conservation strategy).
- Manage for wood production or multiple use, with the same silvicultural techniques as for natural forest management (see Chap. 5).
- Manage as part of an agroforestry system.
- Enrich by planting with species of economic and/or ecological value.

We have already discussed the first three options in this and the preceding chapter. Enrichment planting is discussed in the next section.

6.4.1.1
Enrichment Planting of Degraded and Secondary Forests

Enrichment planting (also known as line-, strip-, gap-, and under-planting) is defined as the introduction of valuable species to degraded forests without the elimination of valuable individuals already present (Weaver 1987; Lamprecht 1990). Enrichment planting has been suggested as a technique for restoration of overexploited and secondary forests as it can increase total tree volume and the economic value of forests (Weaver 1987, 1993; Sips 1993; Ådjers et al. 1995). Enrichment of natural forests after logging may be appropriate in areas where natural regeneration is insufficient or soil characteristics are not conducive to other uses (Ådjers et al. 1995). Enrichment may also include planting fruit trees or other species with commercial or local value. Enrichment planting can be useful as a technique to establish forest species that cannot grow in open plantations because the trees may suffer from continuous direct insolation (Ashton et al. 1995, 1997a, 2001; Montagnini et al. 1997). In Box 6.12 we present results of enrichment plantings in overexploited forests using native species in the subtropical forest of Misiones, Argentina.

Fig. 6.14. Enrichment planting with timber species and with the palm *Euterpe edulis* in Misiones, Argentina. *E. edulis* (for heart of palm) has a shorter harvest age (9–10 years) and has high economic value; therefore it can accelerate and increase investment returns of enrichment plantings. (Photo: F. Montagnini)

Box 6.12

Enrichment planting in overexploited forests in the subtropical forest of Misiones, Argentina (Montagnini et al. 1997)

Line enrichment experiments using native species of commercial value were established in 1988–1990 in overexploited forests in Misiones, Argentina, on public and private lands. The overall goal of enrichment was to improve the forest composition in terms of both quantity and quality of commercially important species so as to allow harvests from short and medium cutting cycles (15–40 years). The species were planted in forests that had been recently harvested using the minimum diameter cutting system, the prevailing cutting method in the region. In the Misiones forests, generally about 40 m^3 ha^{-1} of commercial trees is extracted using this cutting method, and the tree basal area of the residual forests ranges from 10–15 m^2 ha^{-1}. In the region, residual forests with <15 m^2 ha^{-1} are considered overexploited, having relatively slow natural regeneration, and enrichment planting is recommended to accelerate their recovery.

Ten timber species were tested, as well as *Euterpe edulis* (palmito) which can be harvested after 10–12 years for its heart of palm. Enrichment was carried out in lines cut in the forest with east–west orientation to increase light availability for the planted seedlings. All lines were initially 2 m wide and were expanded to 4 or 6 m in the second or third year after planting to increase light incidence for each species, based on general information on the species' light requirements from other experiences in the region. Seedlings were 1 year old, 40–50 cm tall nursery specimens from pots. Seedlings that died were replaced up to the third year. The enrichment lines were weeded two to three times per year during the first 3 years and once or twice per year thereafter as needed.

Four to 7 years after planting, the timber species with greatest mean height and diameter at breast height (DBH) were *Bastardiopsis densiflora*, *Enterolobium contortisiliquum*, *Nectandra lanceolata*, *Ocotea puberula*, and *Peltophorum dubium*. *Cordia trichotoma* and *Balfourodendron riedelianum*, both highly valued timber species, could also be recommended for enrichment despite their relatively slow growth. Labor costs associated with establishment and care of enrichment plantings were similar to other reports of enrichment planting for the region. The incorporation of species with shorter harvest age and high economic value such as the palm *E. edulis* can accelerate and increase investment returns of enrichment plantings (Fig. 6.14).

Because of the relative management complexity of enrichment planting, some authors consider it economically viable only at a small or medium management scale (Ramos and del Amo 1992). The high cost of establishing and maintaining plantings in initial years has been cited as one disadvantage of enrichment techniques (Sips 1993). However, including species of medium and shorter harvest age could improve the economics of this technique, and enrichment planting could play an important role in the recovery of degraded forests (Weaver 1987; Montagnini et al. 1997).

It has also been said that it is difficult to provide optimal light conditions for each species planted within a transect (Weaver 1987) and that introduced species are more exposed to pests than in undisturbed forest (Lamprecht 1990). Most failures can be attributed to the use of species that are not well adapted to this method or to the use of canopy openings that are not conducive to good survival and growth of the species (Weaver 1987, 1993). However, it may be possible to adjust design and management of enrichment lines to avoid some of these problems. For example, light availability can be controlled by changing the width of the transect. Experimentation should include canopy openings of varying sizes, including full open conditions, in order to verify the performances of late-successional species and to develop more fully species selection criteria for enrichment planting. Different topographic positions (valley, ridge) can also influence the successful growth of planted seedlings (Ashton et al. 1997a). Planting some species of the Meliaceae family, which contains many commercially important species, under a partial forest canopy has also resulted in reduced incidence of the shoot borer (*Hypsipyla* spp.) which commonly attacks and severely retards the growth of these species when planted in the open (Newton et al. 1993).

It is important to maintain and regulate the width of enrichment lines so that the requirements of each species are met. Control of surrounding vegetation is the most manageable factor in enrichment systems for secondary forests (Sips 1993). Appropriate line width and light incidence requirements of each species can be determined in controlled experiments. For example, in an enrichment trial on 10-year-old secondary forest in Veracruz, Mexico, the best growth for all species tested occurred at a 68% light transmission with respect to exterior conditions (Ramos and del Amo 1992). In the same experiment, survival was highest in open canopy plantings for the heliophilous species (*Cordia alliodora* and *Swietenia macrophylla*), while *Brosimum alicastrum*, a shade-tolerant, primary forest species, had the lowest survival rate (Ramos and del Amo 1992). In enrichment planting experiments with dipterocarp species in logged secondary forests in Indonesia, a 2-m-wide planting line gave the best results after 2 years (Ådjers et al. 1995). Line plantings do not have to necessarily be in completely straight lines, but can detour around large trees that are in the way.

For mixed dipterocarp forest of middle elevations in Sri Lanka, Ashton et al. (2001) designed a comprehensive set of guidelines that are suited to the specific type of forest, and consider a set of studied understory and canopy species. The guidelines indicate silvicultural treatments that may be necessary in each case, mostly referring to the size of the canopy openings that are needed for each species, and mode of planting (isolated seedlings or in groups or patches), as well as the economic value of each species.

Management of enrichment plantings can be complemented by tending of natural regeneration within the lines. The weeding necessary for initial establishment and maintenance of the enrichment lines also tends to favor regeneration of native species (Montagnini et al. 1997). If regeneration in the enrichment lines is considered together with planted trees, enrichment planting may become a more economically attractive alternative. Therefore, silvicultural treatments should be designed to encourage the establishment and growth of line plantings and at the same time favor natural regeneration, especially of commercially important species (Sips 1993). Once the seedlings are established, the whole forest should be tended throughout, not just along the enrichment lines (Dawkins 1961).

The need for trained personnel and the costs associated with tending may limit the widespread applicability of this approach, especially on a large scale. However, in many regions enrichment planting is a low-risk investment in comparison to other alternatives such as plantation forestry. One way to compensate for the high labor costs of enrichment is to plant species that grow quickly and/or yield highly valued products. In Indonesia, enrichment of depleted dipterocarp forests has become an economically attractive alternative due to increasing lumber prices (Korpelainen et al. 1995). In another example, sensitivity analysis of enrichment plantings in Kalimantan, Indonesia, showed that enrichment of secondary forest with fruit trees, such as *Dialium* spp., *Garcinia* spp., and *Willughbeia* spp., was an economically and ecologically viable alternative (Schulze et al. 1994). In enrichment experiments in the forest of Veracruz, Mexico, Ricker et al. (2000) also showed initial good growth of seedlings of the local fruit trees *Pouteria sapota* and *Diospyros digyna* with optimal canopy openings of 60 and 55%, respectively.

Another option is to combine timber trees with species that produce an earlier profitable harvest, in order to accelerate returns on investment and to make this technique more economically attractive. In Thailand, tamarind (*Tamarindus indica*), which yields a popular fruit, is sometimes interplanted in teak plantations (Jordan et al. 1992). Enrichment adds value to previously logged, low-volume forests by increasing the expected harvestable volume. This added value may prevent their conversion to plantations or other prevailing land uses.

6.4.2
Rehabilitation of Degraded Pasture and Cropland

The initial step in the rehabilitation of degraded land is to identify the most important constraints to productivity, as well as defining the specific land restoration objectives. Objectives can include the recovery of the soil's productive capacity or recovery of biodiversity.

6.4.2.1
Recovery of the Soil's Productive Capacity

Some soils can be recovered through the use of fertilizers, while others that are badly eroded, or that are fully covered by invasive species such as grasses or ferns, need more drastic rehabilitation techniques. There are also situations of extreme degradation where soils cannot be recovered at all. The recovery of the soil's productive capacity is frequently very expensive; thus the techniques used must produce financial returns in order to ensure their adoptability by the local farmers (Montagnini 2002).

Tree plantations and tree–crop combinations represent particularly desirable land-use alternatives for deforested lands with poor natural forest regeneration due to long distance to sources of propagules or intense site degradation. Among the latter, low soil fertility, soil compaction after abandonment from cattle grazing, and invasion by grasses and other aggressive vegetation can be serious obstacles to both forest regeneration and conventional agriculture (Lugo 1988; Nepstad et al. 1991). As the area in degraded lands spreads out, emphasis is increasing on the use of tree species that can grow in such conditions and yield economic products (timber, fuelwood, and other) as well as environmental benefits (soil conservation, watershed protection) (Evans 1992, 1999). An example of the impacts of trees on soil properties and their use in land rehabilitation projects in Latin America is presented in Box 6.13.

Box 6.13

Rehabilitating abandoned lands in the Latin American humid tropics (Montagnini 2001, 2002)

A research program to develop alternatives for the rehabilitation and use of abandoned lands took place from 1987 to 1998 in three humid forest regions of Latin America: the Atlantic lowlands of Costa Rica, the Atlantic rainforest of Bahia in northeastern Brazil, and the subtropical forest of Misiones, northeastern Argentina. In these regions, common situations of rapid deforestation, loss of biodiversity, resource misuse, and land degradation persist. Similar methods were used in the three locations: soil chemistry and nutrient cycling parameters were measured in pure stands of selected indigenous species using adjacent areas free of trees (abandoned agricultural field or pasture, secondary/primary forest) for comparison. The soils under the species, grassy areas free of trees, and adjacent young secondary forest were sampled for soil fertility and nitrogen availability measurements at the three research sites.

In Costa Rica, in just 3 years, soil conditions improved in the tree plantations compared to abandoned pasture. In the top 15 cm, soil nitrogen and organic matter were higher under the trees than in nearby pasture, with values close to those found in adjacent 20-year-old forests. The highest values for soil organic matter, total N, Ca, and P were found under *Vochysia ferruginea*, a species common in mature and secondary forests in the region (Montagnini and Sancho 1990). Subsequent measurements revealed similar trends in the soil parameters in the following years. Based on the standards determined by the Ministry of Agriculture of Costa Rica for soil fertility assessments, the cation levels (Ca, Mg, and K) under most of the tree species were at or above the critical values for agriculture (Montagnini and Sancho 1990). In contrast, the cation levels in the adjacent abandoned pasture soils were too low for the subsistence crops preferred in the region (rice, beans). Soil organic matter also had positive influences on soil physical properties: soil bulk density was lower (i.e., lower compaction) while soil moisture was higher under the trees than in abandoned pasture (Montagnini and Mendelsohn 1996).

In Bahia, Brazil, values of at least five soil parameters under 15 out of the 20 species of the plantations were similar or higher to those found under primary or secondary forest. Several species contributed to increased C and N, including: *Inga affinis*, *Parapiptadenia pterosperma*, *Plathymenia foliolosa* (leguminous, N-fixing species), *Caesalpinia echinata*, *Copaifera luscens* (leguminous, non-N-fixing), *Eschweilera ovata*, and *Pradosia lactescens* (of other families). Others increased soil pH and/or some cations, such as *Copaifera luscens*, *Eschweilera ovata*, *Lecythis pisonis*, and *Licania hypoleuca* (Montagnini et al. 1994, 1995b).

In Misiones, northeastern Argentina, the greatest differences in soil C and N levels under tree species and grass were found under *Bastardiopsis densiflora*, where they were twice those in areas beyond the canopy influence (Fernández et al. 1997). The pH was higher under *Bastardiopsis densiflora* and *Cordia trichotoma*, while the sum of bases (Ca+Mg+K) was highest under *Cordia trichotoma*, *Bastardiopsis densiflora*, and *Enterolobium contortisiliquum*.

Most of the species identified in this research for their positive influence on soil properties are currently being used as components in productive land-use systems such as commercial plantations and agroforestry in each region (Figs. 6.15 and 6.16).

Fig. 6.15. The native tree *Schizolobium amazonicum*, 17 years old, in an arboretum at the Pau Brazil Experiment Station in Porto Seguro, Bahia, Brazil. As mentioned earlier in this chapter, this species can be grown as a nurse tree in reforestation of degraded areas. (Photo: F. Montagnini)

6.4.2.2
Restoration of Areas Invaded by Aggressive Vegetation

In several cases a previously forested area is invaded by aggressive grass, as in the case of *Imperata cylindrica* in Indonesia, *Imperata brasiliensis* in Brazil, *Sacharum spontaneum* in Panama, and *Pennisetum purpureum* in Africa, or by ferns. The competitive advantage of grasses, combined with degraded soils and lack of nutrients, often prevents the germination and initiation of tree seedlings (Kuusipalo et al. 1995). These grassland areas are also often maintained by fires which inhibit natural colonization by tree species (Chapman and Chapman 1996). In these cases, if the area is left for natural regeneration to take place, it may stay in an arrested stage of succession for many years. Often the best alternative in order to restore these areas and revert

Fig. 6.16. *Enterolobium contortisiliquum* (timbó), a nitrogen-fixing tree, in a 9-year-old mixed plantation with other native species on a degraded site in Eldorado, Misiones, Argentina. This species is also used in agroforestry combinations with annual and perennial crops in the region. (Photo: F. Montagnini)

them back to forest or convert them to plantations is planting fast-growing trees, often exotics.

In many cases it is not feasible to plant tree seedlings without first removing the invasive vegetation. In such cases the use of herbicides, fire, or mechanical weeding may be necessary to clear the area and allow tree seedlings to establish successfully. In some cases treatments may involve tillage, or root removal of the invasive plant, as done in experimental settings in Sri Lanka to eliminate the invasive fern *Dicranopteris linearis* (Cohen et al. 1995).

Following treatment to eliminate the invasive vegetation, planting of fast-growing tree species, in many cases exotics, can be a first step to initiate tree cover, suppress the grass, and ameliorate the environment, thus facilitating the establishment of other tree seedlings that may be brought later to restore the original forest, or to give origin to a mixed or a monospecific plantation, depending on the objectives of the restoration project. In Indonesia, successful experiences have been reported on the use of species of *Acacia* to restore sites dominated by *Imperata cylindrica* (Kuusipalo et al. 1995; Otsamo et al. 1999). In the Kibale forest in western Uganda, exotic softwoods (*Pinus caribaea, Cupressus lusitanica*) were planted in the 1960s and 1970s to convert grassland areas into wood fiber-producing sites. In the mid-1990s, tree regeneration was assessed under plantations and compared with adjacent forest. Diversity and abundance of natural woody regeneration were higher under pine than under cypress: the pines apparently provided quick shading which suppressed the grass earlier than the cypress. The composition of the advanced regeneration under plantations differed from that under natural forest. Plantations had similar numbers of species under their canopies compared to natural forests, but these were less valuable species, especially within the commercial classes (Fimbel and Fimbel 1996). However, the application of silvicultural treatments, including enrichment planting, was suggested as an alternative to improve the restoration value of these plantings. Another study at the same site evaluated the role of *Pinus caribaea, Pinus patula* and *Cupressus lusitanica*, and found that although the exotic species had been effective in suppressing the grass, and were an economically attractive alternative, lack of enough regeneration of valuable timber species under their canopies made the exotic plantations not as convenient in the long term, and recommended more research to test the possibility of planting indigenous tree species instead of exotics (Chapman and Chapman 1996).

In areas of the Panama Canal watershed invaded by *Sacharum spontaneum*, PRORENA is testing the effectiveness of native tree species to suppress the grass and restore natural forest, using mixed plantation designs (PRORENA 2003; Fig. 6.17). Prior to planting the trees, a series of treatments to combat the grass include cutting the grass, letting it dry for 1–2 weeks, burning to remove accumulated organic material, spraying herbicide (usually Roundup) after the grass resprouts, and cutting again if necessary (M. Wishnie, pers. comm.).

Other alternatives to initiate ecosystem restoration in areas invaded by aggressive vegetation may involve planting herbaceous species to protect the soil surface, retain soil moisture, improve soil fertility, and retard ground fires. For example, in the central hills of Sri Lanka, four species of nitrogen-fixing leguminous species were used to initiate site amelioration in successionally arrested grasslands: *Calapogum mucunoides, Centrosema pubescens, Desmodium ovalifolium*, and *Cymbopogon nardus* (Ashton et al. 1997b). *D. ovalifolium* was the most successful ground cover at sites with high levels of herbivory, but under conditions of low herbivory the other species were most adequate due to higher biomass production. Establishment of these species was also affected by low soil fertility and pH, and therefore these conditions should be remedied to ensure their success.

Fig. 6.17. An area of the Panama Canal watershed invaded by the exotic grass *Sacharum spontaneum*, where PRORENA is testing the effectivenes of native tree species to suppress the grass and restore natural forest, using mixed plantation designs (see text for details). (Photo: F. Montagnini)

6.4.2.3 Recovery of Biodiversity in Degraded Lands

Plantations of Native or Exotic Species

In abandoned areas, natural forest regeneration is often significantly delayed by physical or biological barriers. The establishment of plantations may overcome some of these barriers by attracting seed dispersal agents into the landscape and by ameliorating local microclimatic conditions within the area, thereby accelerating the recovery of biodiversity (Parrotta 1992; Powers et al. 1997; Lamb 1998).

The establishment of tree plantations in degraded areas may facilitate regeneration of native species that could not otherwise establish in open microsites or in competition with herbaceous species. Several authors report on the role of tree plantations as a catalyst of natural succession in tropical and subtropical sites (Parrotta 1992; Keenan et al. 1999; Parrotta 1999). There are several examples where exotic tree species have been effective in suppressing invasive grass vegetation and recruiting native species under their canopies, as in some of the cases mentioned in the previous section. In Puerto Rico, in the understory of plantations of the exotic tree *Albizia lebbek*, 22 species of trees and shrubs were found, in comparison with just one species in control plots without trees (Parrotta 1992). Even species that have been claimed to have harmful effects on understory plants such as eucalypts have been used with relative success: for example, in highland Ethiopia, 83 plantations of exotic species and adjacent forest stands were analyzed for herbaceous plant cover and biomass, species richness and soil characteristics. The exotic plantation species evaluated included *Eucalyptus grandis*, *E. globulus*, *E saligna*, *Pinus patula*, and *Cupressus lusitanica*.

Fig. 6.18. Mixed plantations with native tree species had relatively high abundance and high numbers of regenerating species in their understory, in comparison with pure plantations, in research plots at La Selva Biological Station in Costa Rica (see text for details). (Photo: F. Montagnini)

Plantations were over 9 years old and had been planted at 2×2 m distance. The richnesss and biomass of herbaceous species in plantations of eucalypts and pines were as high as in natural forest, although most of the species found under plantations were widespread species, mainly weeds or species invading from montane or wooded grassland (Michelsen et al. 1996). Natural regeneration was much less abundant under cypress than under eucalypts or pines; therefore the authors do not recommend planting cypress due to the risk of soil erosion.

However, native species can be as effective or even more effective than exotics in suppressing grass and recruiting native species under their canopies. For example, in north Queensland, Australia, a greater diversity of species was found in the understory of plantations of native species than in plantations of exotic species (Keenan et al. 1999). At La Selva Biological Station, Costa Rica, results of some studies also suggest that tree plantations of indigenous species have a good potential for accelerating the processes leading to recovery of biodiversity in degraded soils (Guariguata et al. 1995; Powers et al. 1997; Carnevale and Montagnini 2002; Cusack and Montagnini 2004; Box 6.14).

Box 6.14

Recovery of biodiversity in pure and mixed plantations with native species in Costa Rica

Mixed plantations may offer a more favorable environment for natural regeneration than pure plantations, due to their multi-strata architecture. Mixed plantations may have a higher variety of environments for seed dispersers and potentially create a greater variety of ecological niches allowing for the establishment of diverse regeneration. At La Selva Biological Station, mixed plantations with native tree species had relatively high abundance and high numbers of regenerating species in their understory, in comparison with pure plantations (Fig. 6.18). Higher plant species richness accumulated under *Vochysia guatemalensis*, *Virola koschnyi*, *Terminalia amazonia*, *Hyeronima alchorneoides*, and *Vochysia ferruginea*, all species commonly planted by farmers in the region. Natural regeneration was higher in understories with low or intermediate light availability. Most of the seeds entering the open pastures were wind-dispersed, while most seeds entering the plantations were bird- or bat-dispersed. This suggests that the plantations facilitate tree regeneration by attracting seed-dispersing birds and bats into the area.

The different species of the plantations created different conditions of shade and litter accumulation, which in turn affected forest regeneration (Carnevale and Montagnini 2002). Competition from grasses is a major factor influencing woody invasion under these plantations. High accumulation of litter on the plantation floor may help diminish grass growth and thus encourage woody invasion under the species' canopies.

High establishment and maintenance costs are potential disadvantages of the use of plantations for accelerating natural regeneration, given the intensive management that is needed, especially during the first 2–3 years (Montagnini et al. 1995a). On the other hand, silvicultural manipulations (thinning, enrichment planting) may increase the economic as well as the ecological value of the regenerating forest. In addition, the value obtained when harvesting the products at the time of thinning or at the final harvest gives an economic incentive for the establishment of these systems.

Other Tree-Based Strategies for Recovery of Biodiversity in Deforested Landscapes
In regions with large agricultural fields that are far from sources of propagules, windbreaks and remnant trees in pastures and agricultural fields may be important reservoirs of native tree species (Guindon 1996; Holl 1998; Harvey and Haber 1999; Harvey 2000; Fig. 6.19). For example, the effects of planted windbreaks on seed deposition patterns were examined on dairy farms in Monteverde, Costa Rica, by Harvey (2000). The windbreaks were planted strips of trees about 5 m wide and 9 m tall and were 7–8 years old at the time of the study. Trees and rows within the windbreaks were spaced at 1.5 m. The most common species were *Cupressus lusitanica*, *Croton niveus*, *Casuarina equisetifolia* (all exotic species), and *Montanoa guatemalensis* (na-

Fig. 6.19. Planted windbreaks on dairy farms in Monteverde, Costa Rica. (Photo: F. Montagnini)

tive). Windbreaks were found to receive significantly greater densities and species richness of seeds of tree and shrub species than pastures. Windbreaks received an average of 39 times as many tree seeds and 67 times as many shrub seeds as pastures. In addition, windbreaks received an average of two times as many tree species and more than two times as many shrub species as pastures. The differences in the seeds entering windbreaks vs. pastures appeared to be due almost entirely to the enhanced activity of birds in windbreaks. Bird-dispersed seeds occurred in greater densities (about 100 times greater) and the number of bird-dispersed species was three times greater in windbreaks than in pastures. The high densities of bird-dispersed seeds within windbreaks suggest that windbreaks increase forest seed recruitment by serving as habitat and/or movement corridors for seed-dispersing birds (Harvey 2000).

Windbreaks may serve as sources of woody colonists if the agricultural lands are later abandoned. Positioning of windbreaks within the landscape may affect seed deposition patterns by influencing the movements of seed-dispersing birds. Tree recruitment may be higher in windbreaks that are connected to forests. Windbreaks could be made more attractive to birds by including native, fruit-producing trees, by increasing their species and structural complexity, and by positioning them between forest patches to facilitate bird movement (Harvey 2000).

Remnant trees in pastures or agricultural fields may play an important role in conserving biodiversity within agricultural systems because they provide habitat and resources that are otherwise absent from agricultural landscapes (Harvey and Haber 1999). For example, in a survey of 237 ha of pastures in Monteverde, Costa Rica, Harvey and Haber (1999) found over 5,000 trees of almost 200 species, with a mean density of 25 trees ha^{-1}. Primary forest trees accounted for over half of the species and over one-third of the individuals. More than 90% of the species were known to provide food for forest birds or other animals. In addition, many of the species were important as sources of timber, firewood, or fence posts for farmers. Reasons for leaving trees in the pastures included using them for shade for cattle, timber, fruits for birds, and fence posts. Results of surveys among farmers suggested that farmers in the region would be receptive to programs promoting the conservation of forest trees in pastures if these programs would fit the particular requirements of shade management for cattle and if they allowed farmers to use a small proportion of their trees for timber, fuelwood, or fence posts. The conservation of pasture trees must be part of larger conservation initiatives that include the conservation of large forest tracts, key habitats, forest fragments, migration routes, and corridors (Harvey and Haber 1999).

6.5
Conclusion

Tree plantations, when well planned and managed, are more productive than natural forests, and thus have a potential to supply the increasing demand for timber worldwide. The total area of plantations worldwide is still relatively small. However, the area and importance of plantations for supplying timber and other tree-based products is increasing. Apart from their productive functions, plantations have important environmental roles, including carbon sequestration and recovery of biodiversity. As wood from native forests becomes more scarce, prices for tropical timbers will rise, and plantations will become more economically attractive.

Agroforestry systems can contribute to accelerate the return of economic investment and serve social and environmental services as well. Several types of agroforestry systems are practiced worldwide, especially in subsistence economies and for restoration of degraded land.

Restoration and rehabilitation of degraded forest lands also are important for re-establishing production and conservation values. The following strategies can be used with good results in restoration of degraded primary and secondary forests:

- enrichment planting;
- improving soil quality through tree plantations;
- ensuring that species that attract seed dispersers are present;
- using mixed species plantations to accelerate natural succession;
- establishing windbreaks, especially with species that offer habitats for animals that disperse seeds.

Plantations, agroforestry systems, and forest restoration can all contribute to solving social and economic problems as well as environmental problems of developing tropical regions by providing economic opportunities for local people. The various ways that natural resource development programs can be implemented, and the strengths and weaknesses of the alternative approaches, are the subjects of the next chapter.

Approaches for Implementing Sustainable Management Techniques

7

7.1 Introduction

The previous two chapters described techniques whereby the basic structure and function of natural tropical forests can be maintained (as in sustainable forest management) or simulated (as in plantations or agroforestry) while at the same time yielding economic profit. While these techniques have great potential for ameliorating the losses and damages caused by conventional logging and forest management, only a very small proportion of tropical forests are managed using these methods. It is not the lack of technical knowledge of sustainable forest management but rather economic and political factors that prevent implementation. Social and environmental benefits of sustainable forest management are intangible, that is, they benefit the global commons through reduced social tensions and decreased environmental degradation. Projects that benefit the forest and the populations that live in and around them produce diffuse results over the long term and therefore garner little political support. In contrast, grandiose projects that generate lots of publicity (but do not include techniques of sustainable forest management), such as the Indonesian Transmigration Program described in Chapter 4, are politically more attractive.

Economic barriers also hinder application of sustainable forest management techniques. Often the dominant view is that the forest should be used to fuel local economic growth, or to establish sovereignty or ownership by harvesting it as quickly and cheaply as possible. The view that tropical forests are a resource that should be used to sustain tribes and communities that have existed in or around the forest for generations is usually considered not economically viable by conventional economic analyses that consider only the present market value of the trees.

In this chapter we ask what political and economic development strategies would be both politically practical and economically effective for achieving implementation of techniques that result in sustainable tropical forest management? This chapter examines two contrasting policy approaches (top-down and bottom-up management) and two contrasting economic approaches (globalization and locally centered development), and analyzes their strengths and weaknesses from the viewpoint of sustainable tropical forest management.

7.2 Top-Down Development

Top-down development derives from traditional colonial exploitation, when European powers laid claims to lands in South America, Africa, and Asia, and "civilized" the native populations. Modern colonialism relies more on exchange of material goods than force to convince undeveloped countries to trade away their natural resources and third-world politicians are often accomplices in this exploitation. However, in both cases, decisions on development plans have been carried out in council chambers, offices or boardrooms located far away from the locale of the resources, but within very close contact of the monetary agencies that finance the development.

In the years following World War II, developed countries initiated aid projects where the stated objective was not so much to obtain the resources of the tropical countries, but rather to improve the "standard of living" of the peoples in the tropical countries. However, the mode of development was still based upon top-down management approaches of economic exploitation with little or no participation from people of the region that was to be developed. Although participatory methods, where the affected people are involved in all aspects of project design, management, and evaluation, were known in the early 20th century for improving labor conditions in developed countries, they were not considered appropriate for aiding undeveloped countries (Castellanet and Jordan 2002). It was reasoned that the resources necessary to be effective in helping lesser developed countries develop economically could only be mustered through agencies that had access to international financing. Further, local communities lacked the expertise to carry out development projects that could integrate their community into the world economy. The complexity of development problems is so great that no single specialist can pretend to know and understand all relevant aspects of the problem. In the top-down or authoritarian approach to development, political or economic authorities decide on a project based on recommendations by outside consultants. The assumption has been that the local residents have neither the knowledge nor the vision to propel themselves into the modern economy and that guidance must be left to the "experts".

However, there are a number of problems with the top-down approach, some of which are illustrated in the case study, "Development and deforestation in the Philippines" (Chap. 4). Perhaps the major problem is that business people who are positioned to take advantage of the infrastructure provided by the development aid benefit while the local people do not. Such business people can be the local elite, national firms operating out of the nation's capital or other large cities, or international corporations. Other beneficiaries are often people from other regions or countries who take advantage of employment or agricultural opportunities in the area being developed. The 1962 (pre-development) FAO report mentioned in the Philippine case study led to the belief that logging the Philippine forests would improve the standard of living of local residents. However, the logging projects, as with many other projects designed and orchestrated by individuals far removed from the practical consequences of the development, failed to consider the impact on the people who lived in or around the areas deforested. Top-down development aid is too often based on global and national economic considerations that have little relation to local social, cultural, and economic realities.

A common complaint against internationally based development programs is that they benefit businesses in the donor countries more than they do in the country being developed. For example, the "second green revolution", which aims to introduce genetically modified crops to developing countries has the potential to bring great profits to the agro-businesses in the USA and Europe, not only through the sale of the genetically modified seed, but also through the chemicals that are necessary to support super-plants (Jordan 2002). However, resistance to introduction of genetically modified crops in many countries illustrates that, although the ultimate goals of authoritarian development efforts may be driven primarily by economic theory, the process through which successful development is achieved at the national or local level is overwhelmingly social and political (Bailey 1996).

Another weakness of the top-down approach is that the expert consultants for many projects are paid by the development agencies or banks and thus the evaluations may be compromised. This is especially true when the consultants have ties to industries that stand to gain from business generated by the development activity. Development loans are particularly good instruments for opening up markets for businesses in the lending country because much of the aid is officially tied to purchasing goods from donor countries (Goldsmith 1996a).

Because of past problems with top-down development, attitudes in international agencies that traditionally operated in this manner have changed in recent years. As the disastrous consequences of unbridled logging became apparent to the world, financial backing for projects such as deforestation in the Philippines has diminished. Nevertheless, top-down, centralized planning often still occurs at the international and national level.

7.2.1 Top-Down Conservation Planning

Top-down planning is not restricted to traditional development projects. Many conservation programs are also top down, and some have had equally unfortunate results. In Africa the establishment of national parks for conservation purposes has sometimes followed a command and control pattern: boundaries were established, everything within them was declared to be legally protected, and a long list of standards was drawn up ranging from a total ban on harvests to the setting of quotas for harvest.

Smouts (2003) describes the case of a huge tri-national conservation project in the Central African Republic (CAR), Congo, and Cameroon. The initial objective was to maintain "pristine" conditions, i.e., to keep large tracts of uninhabited forest lands with intact ecosystems and high densities of large mammals, in particular elephants and gorillas. "We had limited experience in developing or managing national parks", said the director of the project, "particularly in forests with relatively high human population densities or those affected by local logging activities". Unfortunately, there were indigenous Pygmy communities living in this region that was "ideal" for a conservation program. In 1986, the project had to be modified so as to leave the inhabitants the possibility of continuing their hunting and gathering activities. A decade after this initiative was started, there were still similar challenges to reconciling ecosystem and wildlife conservation and the protection of local livelihoods. Human demographic problems, marginalization of Pigmy communities, control of logging activities, and trinational coordination of conservation initiatives were perceived as the major project constraints (Fay 1997). A Sangha River Reserve extending along a 35-km stretch of the Sangha River and incorporating CAR, Congo, and Cameroon protected areas was envisioned as the next step of this conservation initiative. Securing continued funding in the form of endowments was basic for the success of the project: "We have created a system that requires 1.5–2 million US$ per year to sustain". As time goes on, human population pressure increases, logs become more valuable, and the conservation model becomes increasingly strained as funding diminishes (Fay 1997). Top-down management is sustainable only so long as top-down money is available.

The attempt to confer absolute protection to the flora and fauna of regions in Africa has caused other calamities also. For example, the creation of national parks and wildlife preserves in Kenya and Tanzania resulted in the expulsion of Massai stockbreeders. While hunting and logging diminished significantly inside the parks, the problem shifted to the outside: animals were slaughtered as soon as they crossed park boundaries. Kenya is said to have lost, in this way, half of its wildlife outside protected areas in less than 20 years (Norton-

Griffiths 1998). Logging has also shifted to other unprotected and over-exploited areas.

By definition, protected areas restrict uses over a specific area of land. When planned top-down and with only preservation of wildlife as a goal, they deprive communities of land rights they have exercised for generations. In some cases, projects have been accompanied by development and training operations that persuade these communities to change their behavior so they no longer rely solely on hunting and exploiting the forest for subsistence. The heavy social costs of top-down conservation are borne by those who live in protected areas or who have been displaced to the periphery, whereas the benefits go to others. Smouts (2003) names ecotourists, hikers, and photographers who use the parks for recreation as the principal beneficiaries of the parks. The people who benefit most from parks are rarely from the local community. The Kenya-Tanzania park development illustrates how difficult it can be to achieve success with a program that is not accepted by the residents of the impacted regions.

However, there are several examples in which protected areas successfully integrate the local communities in park management and protection. For example, in the Guanacaste Conservation Area in Costa Rica, a World Heritage Site, local inhabitants participate actively in maintenance and management. The protection of Costa Rica's existing habitats is the responsibility of the National Parks Service, which is in charge of the management of 20 National Parks, 8 Wildlife Reserves, and a National Monument. The Guanacaste Conservation Area (GCA) is composed of three National Parks: Santa Rosa, Guanacaste, and Rincon de la Vieja; the Junquillal Bay Wildlife refuge and Recreational Area; and the Horizontes Experimental Forest Station. A series of properties were purchased to form a continuous block of 120,000 terrestrial hectares and 70,000 marine hectares in which there exist approximately 230,000 animal and plant species (65% of the estimated number of species in Costa Rica). The GCA has about 10 years of experience in the search and consolidation of a model based on the reality of tropical ecosystems that can allow conservation of biodiversity through non-destructive uses by society (www.acguanacaste.ac.cr/). In the GCA there are programs in education, research, tourism, reforestation, fire control, and police protection. The total number of staff is about 100, all of them Costa Rican, and four out of every five from the region in and around the conservation area (Allen 2001).

Another successful example of integration of local people in park management is found in the Kibira National Park, Burundi. This national park of an area of 40,000 ha contains Burundi's only montane rain forest. It is a zone rich in both animal and plant diversity. More than three quarters of the water of the country's largest dam – providing more than 50% of the hydroelectric energy consumed – comes from this forest. Thus the park, situated on the Congo-Nile ridge, plays a fundamental role in regulating the hydrological cycle and protecting against soil erosion (Amsallem et al. 2003). In the Kibira National Park, village communities participate in its management through a community conservation plan, which is a pledge of partnership among the people, the administration and conservationists. In addition, a new consultative body, the "local park watchdog community", set up in each of the communes around the park, appears to be a solution to the question of how to involve the people in managing the park.

Parks are just one category of the suite of protected areas in a country. Other categories of protected areas (e.g., wildlife refuges) generally are more flexible in allowing management and in their concerns regarding the livelihoods of local communities. For example, Brazil has a vast array of different types of conservation units that serve different purposes. Areas that are primarily for maintaining natural ecosystems without human presence are classified as "integral protection areas" under the National System of Conservation Units (SNUC). Federal conservation units in this category include national parks, ecological reserves, and biological reserves. By contrast, "sustainable-use areas" promote use of renewable natural resources in the area under management regimes that are intended to sustain production while maintaining the major ecological functions of the natural ecosystem. These include national forests (FLONAs), which are predominantly designed for timber management, and extractive reserves (RESEX), which are intended for management of non-timber forest products such as rubber and Brazil nuts. In the state of Amazonas, a new category called "sustainable development reserve" (RDS) was created in 1996, where local residents zone the designated area into portions for community management of resources such as fish and timber, with a core area that is to remain untouched (Fearnside 2003).

There is a debate about whether humans or economic activities should be allowed in protected areas. One view is that unless the local populations are allowed to benefit from the protected areas in legal ways, they will destroy it in illegal ways. The contrasting view is that parks should address the loss of global biodiversity and global warming which are more critical issues than local impoverishment and conflict. If these environmental problems are not solved, it is argued, the well-being of locals is irrelevant, since humanity will drive itself to extinction. In this train of thought, top-down, mandatory and exclusive protection of tropical forests is the only route to human survival (Terborgh 1999). Debates on this controversial topic are collected in Kramer et al. (1997) and Brandon et al. (1998).

7.3 Bottom-Up Development

Bottom-up development is development in which the inhabitants of an area decide how their resources should be developed or if there is a need to develop them at all. The local people hold community meetings, discuss goals and objectives, formulate plans, and then proceed to implement the plans. The individuals participating in such meetings are the stakeholders, that is, those who have a particular interest in or claim upon the resource in question. For example, in an Amazonian forest area, stakeholders might consist of: indigenous populations who consider the forest a traditional hunting ground; caboclos, or descendants of pioneers who are of mixed race; government-sponsored or independent small immigrant farmers; sawmill operators; loggers; ranch owners who buy-up cleared land; cowboys who work for the ranch owners; fishermen whose rivers flow through the forests; local politicians; government extension workers; and others.

7.3.1 Participatory Action

Community action to develop and/or conserve resources on commonly owned land has a long history. Indigenous peoples treated the land on which they depended as a commonly owned resource, and the practices of many forest dwelling tribes often appear to conserve natural resources (Posey 1982). The concept of the "commons" appears in European history, and Hardin (1968) emphasized the importance of establishing conservation policy over commonly owned resources to prevent their depletion.

Participatory Action Research to promote conservation of commonly owned resources originated from the field of action research, that is, research focused on the resolution of social problems (Lewin 1946). Action research has made significant advances in social innovations (Trist et al. 1963). Because of the apparent improvements in work organization resulting from action research, the concept began to be applied to problems of development in less developed countries. Investigators sometimes played active roles in the process, and as a result it became known as Participatory Action Research (PAR) (Fals-Borda and Rahman 1991), an outgrowth of the Paulo Freire "conscientization" method of the 1960s (Freire 1970).

The basic tenet of PAR is that subordination of the poor in developing countries derives not only from their lack of access to capital but also from their lack of access to education and information. However, they have their own popular knowledge which should be recognized and reinforced through dialogue with modern scientific knowledge. PAR is a way of organizing this dialogue with science and helping people to become conscious of their limitations as well as their potential strengths.

PAR may be limited to evaluation of the development process, or the evaluators may become active stakeholders in the development project itself. The framework for a PAR program is determined by the varying degrees of emphasis given to one or more of the following main characteristics (Jackson and Kassam 1998):

- It empowers communities, organizations, and individuals to analyze and solve their own problems.
- It values the knowledge and experience of local citizens in analyzing their economic, political, social, and cultural reality.
- It uses learning and education to promote reflection and critical analysis by both project participants and development workers.
- Through a learning process, it improves the program and organization in the interests of the beneficiaries.
- It involves the active participation of project beneficiaries, who play a decisive role in the entire evaluation process.
- It promotes the beneficiaries by ownership of a development program.
- It uses a variety of methods, both quantitative and qualitative, to generate knowledge useful in future activities.
- It creates better, more in-depth, and accurate knowledge of the performance and impacts of development projects.

Chambers et al. (1989) applied the concept of PAR to agricultural development. In the following case study, PAR for agricultural development was used in a way that encouraged farmers to adopt methods that help conserve the rain forest in the Amazon region of Brazil.

7.3.2 A Case Study of Participatory Action Research and Development

In the early 1990s, a French non-governmental organization (Groupe de Recherches et d'Echanges Technologiques, GRET) initiated a PAR program for natural resource management in the Altamira region of Pará, in the eastern Amazon region of Brazil.

It was felt that the forest resources were not being used rationally and in many cases wasted. The objective was to provide a forum so that local communities could improve the management of their resources. The program initially was called Laboratório Agro-Ecológico de Transamazônica (see Box 4.7, Chap. 4), but later changed to Programa Agro-ecológico da Transamazônica (PAET) to reflect the broader mission of research plus action. Employees of

GRET (the "researchers") were social scientists and agronomists. In contrast to programs where the evaluators are merely bystanders who chronicle the events as they unfold, PAET members considered themselves to be stakeholders in the conservation of the rain forest.

The first step was to establish a relationship with the farmers' organization in Altamira, the "Movimento Pela Sobrevivência da Transamazônica" (MPST). The farmers' organization was selected in part because it was believed that clearing and burning of forest by farmers was a principal component of deforestation and that farmers continued this practice because of lack of knowledge about alternatives when crop productivity in an area declined. The researchers assumed that farmers were interested in better resource management and therefore:

- They would facilitate the research.
- They would help the research team produce more client-oriented results.
- They would facilitate dissemination of results at the regional level.
- They would represent the farmers in negotiations with regional and national governments.
- They would actively promote activities at the farm and community level that would improve resource conservation and management.

A further assumption was that it was possible to define a common strategy between the farmers' organization and the research team through continuous negotiation and debate. If the MPST and the researchers had a common interest in improving management of natural resources, it should be possible to reach a common understanding or diagnosis of the problem to be treated. Based on this common objective, it would not be too difficult to devise a common strategy, and to negotiate the priorities within this strategy. Negotiating a strategy was seen as the result of a process of improving the communication between the various actors and their mutual understanding of each other's views and positions.

Planning of the action research was to be evolutionary and cyclical, based on regular evaluations and adjustments in the program (Rhoades and Booth 1982). This "Basic Action Research Cycle" was based on the following steps:

1. The researchers carry out a preliminary diagnosis of the condition of natural resources in the region.
2. They present their findings at an annual regional seminar.
3. The farmers respond with their views.
4. Priority problems are selected through negotiation.
5. Planning activities are carried out. Activities can be basic research, applied research, training, and technical assistance to the organizations.

6. Activities are carried out with specific target groups.
7. At the end of the year, technical and scientific results are presented to the target groups, and results evaluated from the farmers' point of view.
8. The initial diagnosis of the researchers is reviewed in light of the new evidence.
9. At the annual seminar, progress of the general program is evaluated by representatives from all organizations.
10. Back to step 1.

A crucial difference between this action research cycle and programs in which a grass roots movement is merely evaluated is the feedback loop. The yearly evaluation allows participants to see what is working and elaborate on those initiatives and to discard or revamp those initiatives that are not satisfactory.

Castellanet and Jordan (2002) published an account of the first 5 years of this program, and focused particularly on two communities in the Altamira region, one where the program was judged to be successful and another lacking positive results. In the following section, we present a synopsis of the program in each community and an analysis of the reasons for success or failure.

7.3.2.1
Case I: Uruará: Where PAR Failed

The Problem

The selective logging of mahogany and cedar carried out in the 1970s resulted in the opening of roads that extended 100 km or more from the village of Uruará. These roads facilitated access to remote forests that were quickly occupied by families who cleared the area for crops or pastures. Many socio-economic problems arose from the disorderly occupation of opened forest. Maintenance of the dirt roads was difficult. Neither the municipal government nor the loggers wanted to take responsibility. Due to the resultant high costs of transporting products, profits for farmers who lived along these feeder roads were very low. In addition, access to social services such as schools and medical clinics was also severely limited. Isolation made it difficult for the community associations to meet regularly. Added to these was the problem of fire in areas that have been recently logged.

Another important problem in the region was the loggers' invasion of indigenous peoples' reserves. Contacts with the "white men" frequently resulted in destruction of the indigenous tribes, especially due to alcoholism. Indigenous people gained cash by selling trees, often at prices lower than those offered to farmers. Selective extraction also caused significant impact on game, which is an important food source for the indigenous tribes.

A conference of stakeholders on the forestry issues was held in March 1995 at Uruará. The most important proposals that emerged were:

- The creation of a natural reserve and a pilot municipal forest, to be managed sustainably by the community.
- The enhancement of the value of wood by creating cabinet-making workshops.
- Sale of trees at a value based on their volume, not at a price bargained for by the logging companies with individual farmers.

The idea of creating a natural reserve and municipal forest was supported by the logging companies, even though these proposals were not favorable for them. Perhaps they believed that the decision was just a façade and would never be truly implemented, or perhaps they thought that by agreeing to the idea, they could more easily open a route to a river port in the north of the municipality. In any case, 2 years later, nothing had happened with regard to a municipal forest and the logging companies had opened their route to the north, even though they had no public support for the project.

There was initial enthusiasm among the farmers for processing wood, not only for cabinets, but also for their own houses and furniture. The problem was that sawing logs with a medium-sized mill and then selling the boards would give rise to problems such as financial control, marketing, and other management problems. Such an enterprise would require an industrial structure which is difficult to establish and maintain. As a consequence, the project was put on hold. Meanwhile, consultants from the Brazilian government recommended that a project for enhancing the value of wood should be given to a national agency, "Fundação para o Desenvolvimento do Município de Uruará" (FUNDASUR). Once the farmers learned of this recommendation they dropped all interest in the PAR approach. The FUNDASUR initiative was sponsored in part by an international bank and held much higher promise for bigger and newer equipment than the PAR program. The FUNDASUR project, however, was strictly top down, and all decisions were made by this agency.

The issue of selling price for logs also did not reach a satisfactory conclusion. When the logging companies bargained with each settler individually they risked not being able to purchase the wood from certain lots, even though they had to open roads throughout the territory. Farmers who refused to sell when the roads were opened later benefited by selling their trees at a higher price to other logging companies who had not paid anything for opening the roads.

Lessons from Uruará

Although a number of innovative proposals had been formulated, it became evident later that the proposals had been negotiated between different influential local groups and did not generally ensue from PAET's studies or discussions. The researchers concluded that they had been used by local leaders to legitimize their own proposals and that the most powerful user groups ended up benefiting the most, either directly from the resources that were mobilized or indirectly through enhancement of their political image.

The major lesson of the Uruará experience was that the dynamics of local planning could not be understood without also analyzing the objectives and strategies of the numerous groups involved: technicians, politicians, tradesmen, churchmen, local teachers, and public and private organizations at regional and national levels. To establish the balance in favor or the numerous less powerful actors, it seems necessary to concentrate on helping these groups strengthen their own organizations and develop their own proposals before they reach the negotiating table.

Lastly, on the frontier such as exists around Uruará, it is not always possible to achieve good conflict resolution through negotiation and discussion of all concerned parties. Once the research team understood this reality, an option was to forget the idea of local planning and to concentrate on the reinforcement of democracy and the establishment of reliable law enforcement.

7.3.2.2
Case II: Porto de Moz: Where PAR Succeeded

Porto de Moz is a district with a more traditional population than Uruará. It is located along one of the main rivers and many inhabitants have lived there for generations. The main activity until the 1960s was rubber tapping, Brazil nut collecting, and fishing. By the 1990s, however, logging represented approximately 60% of the total community income. In 1995, LAET was approached by the leaders of the Porto de Moz district for assistance in organizing a seminar on the future of wood and fish in their district. Based on their previous experience in Uruará, the researchers adopted a different strategy for participatory planning. Rather than inviting all the local stakeholders at the beginning, priority would be given to a direct dialogue with the local rural people's organizations, since these were the people who most needed empowerment.

Activity began with a rapid natural resource appraisal. The research team quickly established a picture of the district's social, economic, and ecological situation. Major themes were (1) that large quantities of trees were harvested and exported outside of the district, and (2) that returns to the district were low compared to returns that could be obtained for sawed lumber. It was estimated that all nearby forests would be exhausted within 10–15 years if nothing was done to curb indiscriminant logging. Community representatives made various statements about the links between forest exploitation and reduced abundance of fish. Fishing was becoming increasingly difficult due to severe competition from commercial fishing boats.

After the seminar, a committee was formed to develop a natural resources management program. Within 2 years, the effort had the following results:

- There was a rapid spread of community rules restricting fishing in the rivers that effectively gained control of commercial fishing in the area.
- Community forest reserves were planned and demarcated.
- A program of environmental awareness was organized.
- Records were obtained from the Public Land Office when the local representative was thought to be corrupt.

One of the reasons that PAR was more successful in Porto de Moz than in Uruará was that the local communities were more structured and the people's link with their land was stronger, since many had lived there for several generations. Another important aspect of the experience was that some of the results were obtained through direct contacts of the local organization with the national administration. This result confirmed the thesis (Sawyer 1990) that natural resource management in the Amazon can be better obtained through direct cooperation of the local organizations with national administrations, thus bypassing the local elite, who are likely to block all initiatives that go against their immediate economic interests.

Partnership with the Farmers' Organization

The initial assumptions on the role of the farmers' organization in Uruará and Porto de Moz were only partly confirmed. For example, the farmers' organization did have an interest in sustainable development and better management of natural resources. However, the establishment of a common strategy was not achieved. The initial model of strategy-building through the improvement of communication between farmers and researchers was found to be inappropriate. It was not possible to conclude that the choice of the farmers' organization was the most appropriate for PAR on natural resource management on the frontier. Researchers cannot expect that the representatives of farmers' organizations will necessarily state their own priorities and strategies at the beginning of the cooperation. Researchers interested in establishing partnerships with local organizations should bear in mind that from the outset, efforts must be made to identify each organization's critical fields of interest. For example, in the cases discussed here, factors that were important to the farmers' organization but were never explained to the researchers were (1) desire for recognition by the public and by local and national institutions, and (2) backing by the local farmers. Until the researchers finally understood this, they were viewed by farmers as competitors for local and national recognition.

7.4 Community Forestry

Community forestry has the potential to combine the strengths of both the top-down and bottom-up approaches. The top-down strength would be the ability to mobilize large amounts of capital, and to enforce policies and decisions regarding land use. The bottom-up strength would be local participation in decisions, thereby ensuring that local culture and environment could be included.

The idea that forestry should aim at bringing a bigger share of forest benefits to local people is embedded in the concept of social forestry. At FAO, it is called "forestry for local community development" (Arnold 2003). Social forestry is defined as any situation that closely involves local people in forestry activities, for which people assume responsibilities, and from which they derive a direct benefit through their own efforts. Because of their complexity, many environmentally sound forest management practices are best suited to small farmers, agricultural cooperatives, or community forest users, rather than to larger farms or larger-scale entrepreneurs (Montagnini et al. 2002). Community forestry has the following general characteristics:

- It involves communities in a continuous decision-making process.
- It involves communities in forestry activities.
- It involves indigenous peoples and farming communities in resource management.
- It values the forest resources according to the cultural heritage of people.
- It is adapted to the farmers' realities and their methodologies so as not to be in conflict with prevailing social and economic processes.

Community forestry can operate at different levels, from a subsistence level in isolated or marginal communal groups with minimum market development, to well-defined and organized groups with good access to technical information where production is oriented to certified markets. Examples of community forestry projects at different levels are presented in the following case studies (Boxes 7.1 and 7.2).

Box 7.1

Community forestry concessions in Petén, Guatemala (Cuellar 2002; Cusack et al. 2002; Montagnini, pers. observ.)

Community forestry concessions in Petén, Guatemala, are excellent examples of communal involvement in forestry project planning and execution. They are also an interesting example of how sustainable forestry practices can coexist with forest protection and preservation of cultural values.

The National System of Protected Areas in Guatemala includes 3 million - ha or 28% of the country. Of the protected areas, 80% is found in the Petén department, which is also the forest reserve of the country. The history of Petén dates back to the years 200–900 b.c. with the predominance of the Maya civilization. In colonial times and following the decline of the Maya civilization, Petén had a very low population and was ignored by the principal social, political, and commercial centers of the region. From the late 1800s until the mid-1950s, the extraction of gum from "chicle" (*Manilkara zapota*) and the harvest of mahogany (*Swietenia macrophylla*) were the main commercial activities. Colonization of the region was complete by 1960. Population has grown significantly over this period, from about 20,000 people in 1960 to over half a million by 2000. This population growth resulted in a loss of forest cover from 90% to less than 50%. Forest was replaced by shifting agriculture. Contracts for forest exploitation were granted to logging companies in 1970–1980 focusing on industrial extraction of mahogany and cedar (*Cedrela* spp.). These logging activities were carried out without using forest management plans and resulted in a substantial decrease in tree populations of both commercial species in the concession areas.

CONAP (the National Council of Protected Areas of Guatemala) was founded in 1989, and in 1990 the Mayan Biosphere Reserve (MBR) was designated, and covered 2.1 million ha (Fig. 7.1). CONAP was assigned the administration of the MBR. The restrictions posed by CONAP on the use of natural resources provoked a set of social conflicts and reaction from the population of Petén against CONAP. This motivated illegal extraction of forest resources (timber, palms, fauna), increase in immigration, and an advance in the agricultural frontier.

However, a master plan was started in 1992 for the management of the MBR with the objectives of protecting biodiversity and promoting sustainable use of natural resources. A zoning system delineated three sections: a nucleus (national parks, no use), a multiple-use zone (MUZ), and a buffer zone. Community concessions were assigned for management in the MUZ in management units (MU) that allowed use of natural resources and transformed the communities into allies of CONAP with respect to resource protection. In the community concessions, forest management, extraction of non-timber forest products, agricultural activities, and tourism are all supported. There are also industrial concessions where only timber extraction is allowed. In both types of concessions forest certification has to be obtained and maintained for the duration of the contract (25 years). The assignment of a concession is in three steps: definition of the MU, public offering, and granting the concession.

The first concession was granted in 1994. There was very high demand from the communities to obtain the concessions. As of 2002, a total of 16 forestry concessions had been granted (14 community concessions and two industrial concessions) over almost 600,000 ha in the MUZ. Forest cover in the concessions is over 98%, and the beneficiaries are about 7,000 people in about 1,300 families. Funding has come from USAID, and technical assistance has been provided by CATIE, local community groups, and non-government organizations (NGOs).

As of 2002, some of the main accomplishments were: greater control of forest fires; zoning and land management; better control of immigration and advance of the agricultural frontier; improved control of illegal extraction of natural and archaeological resources; higher employment; higher income and higher minimum salary in relation to the rest of the country; development of infrastructure; change in people's attitude from individualism to community organization; a more positive attitude towards the forest; and forest certification (317,000 ha certified under communal groups). However, some limitations were also identified. For example, there were some voids in technical information regarding forest management; too few commercial tree species were considered in forest management plans; there was a lack of effective organization; and conflicts of interest were apparent among people in the communities. Among their priorities, the communities agreed that they needed to increase their knowledge of techniques of sustainable forest management; increase their understanding of management of natural regeneration (silvicultural treatments); learn more about timber properties of commercial species; and assess the financial feasibility of the concessions.

Forest management can vary among the concessions. For example, with the Carmelita community forestry concession (granted in 1997), the General Management Plan indicated that the cutting cycle was 40 years with an annual area for cutting of 400 ha. Minimum diameters for cutting are 55 cm for primary species and 45 cm for lesser known species. Only about 1.5–2.5 trees of the most valuable species (mahogany) are extracted per hectare. (This is a very low volume for a profitable operation.) Most of the production is sawn timber that is exported to the USA. Profits from forest management tend to compensate the low profits that are obtained from the other extractive activities such as chicle, ornamental plants, and pepper (Fig. 7.2). The Carmelita Cooperative certified the MU with Smartwood in 1999 for 5 years. Certification has to be renewed for the total duration of the concession contract (25 years, renewable). Certification is expensive because qualified personnel have to be contracted. An alternative is to train local personnel who can help with the certification process. Other alternatives to increase the economic benefits of forest management involve using lesser known timber species apart from mahoganies and cedars, and selling the environmental services of forests.

Overall, the success of community forestry in Guatemala can be measured using economic parameters. For example, community foresters in Petén earn an income that is generally better than other professions in Guatemala (S. Ortiz in Cusack et al. 2002). It is also important to notice that with the increase in community forestry, deforestation rates in the country have dropped from 25,000 ha/year in 1989–1994 to 1,000 ha/year in 1995–2002. In addition, fire events have been fewer in locally managed forest concessions as compared with forested areas that are not in concessions and even in comparison with protected areas.

Fig. 7.1. The Mayan Biosphere Reserve, Petén, Guatemala. (Photo: F. Montagnini)

Fig. 7.2. Drying pepper that was collected in the forest, one of the extractive activities practiced by members of the Carmelita community forestry concession in Petén, Guatemala. (Photo: F. Montagnini)

Box 7.2

Community-managed land-use systems at Coope-San Juan, Costa Rica (Montagnini et al. 2002)

Some farmers' cooperatives in Costa Rica manage their natural forests for ecotourism and non-timber forest products, and carry out other productive activities, including conventional agriculture, on other portions of their land. For example, the Coope-San Juan Agricultural Cooperative, in Aguas Zarcas, NE Costa Rica, has 16 members (11 men and 5 women) who, along with their families, form a community of about 56 people. They collectively own 400 ha of land, half of which is covered with primary forest. They are keeping their forest intact, have marked trails for tourism, and are expecting to obtain payment for environmental services from the legal system currently in operation in Costa Rica. On their agricultural land they keep a dairy farm and sell the milk locally. They also grow cocoa and plantains commercially.

Additionally, they manage non-timber forest species for sale, including a medicinal plant, raicilla or ipecacuana (*Cephaelis ipecacuana*), which they grow in the natural forest understory (Fig. 7.3). There is an export market for ipecacuana in Germany and Belgium. In addition, they have been reforesting portions of degraded agricultural land since 1987 with native and exotic species, often using mixed-species planting schemes (Fig. 7.4).

Cooperatives such as Coope-San Juan are a promising model for more environmentally friendly forestry systems at small to medium scales. For these systems to be successful there may be a need for initial economic incentives and training programs in cooperative management and administration, as well as in the technical aspects of sustainable forest and agricultural management techniques.

Fig. 7.3. In the Coope-San Juan Agricultural Cooperative, Costa Rica, one of the productive activities involves growing a medicinal plant, raicilla or ipecacuana (*Cephaelis ipecacuana*), in the natural forest understory, for export to European markets. (Photo: F. Montagnini)

Fig. 7.4. In the Coope-San Juan Agricultural Cooperative, Costa Rica, farmers are reforesting portions of degraded agricultural land with native species, such as *Vochysia guatemalensis*. (Photo: F. Montagnini)

There are other classical examples of community forestry experiences in Mexico, where forest management is practiced in a communal fashion in the ejidos. The beginnings of the system in Mexico can be traced back to the Mexican revolution of 1910, with the dissolution of the haciendas owned by the elites and the creation of a tenure property system for agricultural workers and indigenous peoples. As part of the Constitution of 1917, a system of ejidos was created whereby a group of farmers owned and managed their land communally. A significant portion of land was also maintained as communal agrarian plots that have existed since precolonial times (Thoms and Betters 1998; Bray 2003). Other examples of forest management include extractive reserves in the Brazilian Amazon (C. Peters in Cusack et al. 2002); farmers' cooperatives to grow coffee under the shade of valuable timber species in El Salvador (Figs. 7.5 and 7.6); as well as other examples of small-scale forest management in community forestry projects in Asia and in Africa. Often such projects involve agroforestry practices as well as forest management.

Although community forestry projects are highly susceptible to failure due to lack of funding, inadequate educational levels among the practitioners, and a potential for corruption and mismanagement, evidence from a number of successful experiences indicates that they can facilitate sustainable forestry. Where

Fig. 7.5. In farms that are part of a cooperative in Tacuba, El Salvador, farmers grow diversified coffee, using a dense overstory of valuable ornamental, fruit, and timber species such as *Cedrela odorata*. (Photo: F. Montagnini)

aid money is available from international agencies chances of success may be better. For example, the World Bank is currently funding a project in Ecuador through the Maquipucuna Foundation to protect a forest corridor from the Andes to the Choco region on the Pacific coast (http://www.arches.uga.edu/~maqui/). The thrust of the project is to encourage local communities to manage their lands in ways that do not destroy the forest cover. Successful community forestry projects can actually provide the most viable hope for both environmentally and socially sustainable forestry in many regions of the tropics, especially when they combine the advantages of top-down and bottom-up approaches.

Fig. 7.6. In many community forestry and agroforestry projects in El Salvador, farmers process the wood on site using a hand saw. The wood is from valuable timber species that they use for shade of coffee. (Photo: F. Montagnini)

7.5 Globalization

Globalization results in an increase in international trade. It is beneficial to corporations that can take advantage of international opportunities as the free market directs them to countries and regions where commodities can be produced at the lowest cost. It also seems beneficial, at least at first, to countries where commodities are produced, because they are then able to use the foreign exchange to subsidize "modernization", which promises to improve the lives of citizens. With time, however, the benefits disappear. The resource may become depleted, or the international company that is buying the resource finds another country that will supply it more cheaply. By this time, the producer nation is accustomed to a so-called higher standard of living and the government cannot easily survive with reduced international income. The next step is to go into debt to international lending banks and agencies, which brings even greater pressure to liquidate more quickly the remaining natural resources. This scenario often is the cause of tropical deforestation, as discussed in Chapter 4.

However, globalization is not necessarily destructive to all aspects of the economy. Within the community of mainstream economists there is considerable debate about the pros and cons of globalization:

"The most pressing moral, political and economic issue of our time is third-world poverty. Globalization through technology offers the best hope of remedying the problem. The advances achieved in computing and telecommunications in the West offer enormous opportunities for raising living standards in the third world. New technologies promise not just big improvements in local efficiency, but also the potentially bigger gains that flow from an infinitely denser network of connections, electronic and otherwise, with the developed world" (The Economist 2000).

However, in the same journal, Wade (2001) wrote:

"Technological change and financial liberalization result in a disproportionately fast increase in the number of households at the extreme rich end, without shrinking the distribution at the poor end. Population growth, meanwhile, adds disproportionately to numbers at the poor end. The prices of industrial goods and services exported from high-income countries are increasing faster than the prices of goods and services exported by low-income countries. As a result, the gap between the richest 10% of the world's population and the poorest 10% is widening".

Rosenberg (2002) has made the argument that globalization has failed the world's poor, but it is not free trade that has hurt them – it is that the system is manipulated. She feels that the architects of globalization are right, that international economic integration is not only good for the poor, it is essential. To embrace self-sufficiency or to criticize growth is to glamorize poverty. No nation has ever developed over the long term without trade. For example, since the mid-1970s, Japan, Korea, Taiwan, and China have lifted 300 million people out of poverty, chiefly through trade. However, she also points out that those who protest globalization are also right. No nation has ever developed over the long term under the rules being imposed today on Third World countries by the institutions controlling globalization. The United States, Germany, France, and Japan all became wealthy and powerful nations behind the barriers of protectionism. East Asia built its export industry by protecting its markets and banks from foreign competition and requiring investors to buy local products and build local know-how. These are all practices discouraged or made illegal by the rules of trade today.

There are claims that globalization of agriculture has helped third-world farmers and contributed to the alleviation of hunger. Southeast Asia has seen impressive gains in food production as a result of the Green Revolution. During two decades of Green Revolution advances (1970–1990), figures from the United Nations showed that the total food available per person in the world rose by 11%, while the estimated number of people suffering from hunger fell from 942 to 786 million, a 16% decrease (Rosset and Mittal 2001). Despite the Green Revolution, World Health Organization data showed that even now, one third of the world's children suffer from malnutrition (De Onis et al. 2001). Hosington et al. (1999) cited the need for genetic engineering "to feed a world population growing by up to 160 people per minute, with more than 90 percent of them in developing countries". Mann (1997) predicts that through the work of plant breeders, crop physiologists, and botanical geneticists, humankind ultimately will be able to feed itself, but only if the world engages in a "gigantic, multiyear, multibillion-dollar scientific effort that emphasizes genetic engineering".

However, not all agree. The impact of globalization on agriculture is a particular worry for environmentalists. "Expanded economic growth and global development", writes Goldsmith (1996b) "cannot be achieved without an immense overuse of resources, a fierce assault on remaining species of flora and fauna, the creation of toxic wastelands and seas, and the degradation of the planet's natural ability to function in a healthy way. The idea, promoted in corporate circles, that first we must make countries wealthy through development and then take care of the environment is high cynicism, since development does not produce wealth, save for a few people; the wealth that is produced is rarely spent on environmental programs; and anyway, by the time the theoretical wealth is generated, life will be unlivable".

Herman Daly, a former staff member of the World Bank, has written scathing denunciations of globalization, from the viewpoint of an insider with privileged access to the results of development projects aimed at integrating the global economy. "Globalization", he writes, "risks serious consequences. They include standards-lowering competition to externalize social and environmental costs with the goal of achievement of a competitive advantage". This results in a race to the bottom as far as equity in income and environmental standards are concerned. Globalization also risks increased tolerance of mergers and monopoly power in domestic markets so that corporations become big enough to compete internationally. Monopoly power in agriculture results from trade-related intellectual property rights, such as the need for third-world farmers to pay royalties to international agri-businesses when the farmers plant genetically modified seeds. These practices put the farmer in debt to these corporations (Daly 2001).

7.5.1 Globalization and Forest Resources

While there appear to be equally strong arguments for and against globalization in industry and agriculture, the impact of an uninhibited market economy on tropical forests appears almost entirely negative. International lumber companies will buy timber first from countries that do not regulate or control logging, and therefore can sell it more cheaply. In effect, unrestricted trade imposes lower standards (Daly 1996). Competition among tropical countries for contracts with international companies results in give-away prices for a logging concession (Gillis and Repetto 1988). Since national governments receive little economic benefits from intact forests, there is little reluctance about letting the forests go for any meager price, since a low price is better than none at all.

Another negative aspect is that most forests in tropical countries are public lands and are controlled by the national government (Repetto 1988). The politicians grant logging concessions and the national government receives the proceeds from logging contracts. If the politicians are honest, the proceeds usually go towards payment of a national debt. Very little, if any, benefit goes to the people that live in and around the forest, or for environmental protection. The claim that deforestation creates local employment is bogus. The percentage of local inhabitants that are hired by logging companies is very small and the work is usually temporary. After the forest is gone, there are little or no resources to which these people can turn.

7.5.2 Case Study of Globalization

The following case study shows how globalization impacts the economy and social relations of local rice producers in the Sahel, Africa (Box 7.3). While the focus is on an agricultural commodity, there is a link to deforestation in that the increase in rice production needed to compete in the global economy required deforestation for new plantings.

Box 7.3

Rice cultivation in the Sahel, Africa (Ba and Crousse 1985; Halstead and O'Shea 1989; Park 1992; McIntosh 1993; Zeng 2003)

Since the late 1960s, the Sahel, a semi-arid region in West Africa between the Sahara Desert and the Guinea coast rain forest, has experienced a drought of unprecedented severity that is leading to desertification. The drought has been caused, in part, by overgrazing and conversion of woodland to agriculture. Both of these processes tend to increase albedo (reflectivity of the earth's surface), thereby reducing moisture supply to the atmosphere. As a result, there is less precipitation and even less favorable conditions for vegetation (Zeng 2003).

Agricultural expansion has been one of the goals of development projects in the Sahel region. It has been driven by the need to increase the national market economy through production of rice. The development programs usually required cultivators to plant the Asian variety of rice, *Oryza sativa*, which is more marketable and has a higher yield than indigenous varieties. However, construction of dikes was required in order to ensure a steady supply of irrigation water required by the Asian rice.

Indigenous varieties of rice have been cultivated for centuries by the Marka, a local ethnic group. They make complex and sophisticated decisions about when to plant and what varieties to plant. Their decisions are influenced by environmental clues – different varieties of rice have different vegetative periods, different adaptations to various flood depths, flood timing, pH tolerance, and resistances to fish predation. Different varieties are sown at different time intervals on different soil types. The knowledge that the Marka possessed about rice and its cultivation was secret and had been developed over a long period of time. It was a means of maintaining a specific ethnic identity. Social relations with other groups were instituted as buffering mechanisms against potential bad times, allowing trade to occur without the necessity of immediate equal compensation. This buffering was useful, for example, with the Bozo fishermen, who reciprocated labor, goods, and services with the Marka. The buffering was beneficial to both groups, because weather that favored one group disfavored another.

Another important aspect of the Marka system of sustainable rice production was prioritized tenure on property held in common with the entire ethnic group. A hierarchical system prioritized access to land and the rules regulating access to common property were encoded in Islamic law. Prioritized access ensured that those with the specialized knowledge were those that made decisions on varieties of rice to be planted as well as on the timing of the planting.

Ensuring sustainability of rice production required a considerable understanding of the social system and its deep ties to the environment through culturally mediated and specialized relationships. Knowledge of the physical needs of a particular crop was not enough to produce consistent quantities in a sustainable manner. Social needs were also important. Farmers made decisions based on variables that seemed unscientific because the farmers were considering these variables from a different temporal and spatial scale than normally understood in the developed world. To understand the farmers' decisions, one had to understand the evolutionary nature of secret knowledge and inter-group relations which functioned together as part of a subsistence system and which buffered the system against environmental and political variability.

As a result of development projects, prioritized tenure on commonly held lands was eliminated in many areas and equal access was gained by those without the secret environmental and social knowledge. The traditional allocation system, built on the recognition of natural variables, was replaced by a system organized to suit the demands of a market economy.

Adoption of commercial systems of rice production has resulted in ecological deterioration. In Senegal, 40,000 ha put under irrigation for rice is now degraded, as inexperienced people quickly erected poorly built irrigation structures in order to satisfy government requirements for establishing tenure. Polders (diked areas) constructed to control water flow were not flexible enough in times of drought. Polders also affect fishing, as changes in the flow of the river and the displacement of water through polders affected fish breeding and feeding. The transition to a market economy ignored the nature of the Sahelian climate and soils and deprived traditional Marka groups of their ability to flexibly respond to environmental variation.

As a result of development, both the Marka and Bozo have formed relationships with the state government and development agencies. As individuals are drawn into the market economy and group identity becomes less important, myths have been changed or are no longer told at all. Traditional knowledge of cultivation and fishing is almost lost. Rice production increased in the short term due to artificial subsidies. While this resulted in great profits for groups well integrated in the national and international economy, it has done little for the betterment of the local people. Long-term sustainability of rice production is lost as native species and the knowledge of their cultivation is abandoned.

7.6 Locally Centered Development and Integrated Natural Resource Management (INRM)

Globalization seeks to eliminate all international barriers to trade, so that commodities can be produced more efficiently. The result is specialization in the production of the commodity that a country or region can produce most cheaply. Locally centered development takes another approach. Its emphasis is on developing local economies with a view to including things that are not valued in the marketplace, such as the interdependence among groups like the Marka and the Boza, illustrated in the Sahel case study. It emphasizes other values in life besides economic efficiency. A foremost value that it seeks to protect is diversity, even though production of a broad diversity of crops (or of crop varieties in the case of the Marka) may reduce local monetary income. Advocates of locally centered development argue that the higher monetary costs of promoting diversity in production are compensated for by other factors: the ecological benefits of diversity, such as reduced pest and disease pressure, and higher production resulting from ecological complementarity; the economic benefits of having a variety of products, so that when one crop fails or has a low market value the producer has alternative crops to provide income; the cultural benefits of having a variety of behavioral approaches to economic and production problems. Localization also improves morale, because community members have a greater influence on the direction of the project.

Recently, many of these concepts have been embedded in what is called "Integrated Natural Resource Management" (INRM), a concept that was advanced by personnel working on development projects from centers that are part of the Consultative Group for International Agricultural Research (CGIAR; CGIAR-

INRM Group 1999). The concept has been applied successfully to a number of development projects worldwide (Hagmann et al. 2003; Sayer and Campbell 2003). The basic strategy to strengthen the adaptive capacity of the integrated natural resource management system at the local level is:

1. Local organizational development, which serves to strengthen the collective capacity of local groups, institutions, and organizations for self-organization, collective action, negotiation of their interests, and conflict management, as well as their articulation and bargaining power vis-à-vis authorities, service providers, and policy makers.
2. To enhance farmers' capacity to adapt and develop new and appropriate innovations by encouraging them to learn through experimentation, building on their own knowledge and practices, and blending them with new ideas in an action learning mode. Usually these are agricultural technologies and practices, but they also address social, organizational, and economical innovations.
3. To enhance collective learning through action and social learning, facilitation of self-reflection, sharing knowledge, and networking.
4. To negotiate the management of natural resources and related services, policies, etc. through stakeholder platforms of communities, service providers, and other key players.

This core strategy is implemented through a variety of concepts, methodologies, and supporting strategies. The INRM process is mainly guided by the vision and values to which the intervening and facilitating agents, as well as the communities, agree and subscribe. These core values are (Hagmann et al. 2003):

- full ownership of the process by the community and control over their own resources;
- self-reliance of local communities;
- self-organization, sharing, and cooperation;
- inclusivity of all stakeholders and groups;
- equal partnership among farmers, researchers, and extension agents, who can all learn from each other and contribute their knowledge and skills;
- equitable and sustainable development through negotiation of interests among these groups and by providing space for the poor and marginalized in collective decision making; and
- natural resource conservation as part of the generation contract.

INRM work began in Zimbabwe in 1988 as part of a collaborative program between the national agricultural extension services (AGRITEX), German development cooperation (GTZ), and the Food Security Project of Intermediate Technology of Zimbabwe (ITZ). The program started with a technical research focus on soil and water conservation in the semi-arid areas of southern Zimbabwe, but it gradually integrated more technical and social elements of the rural livelihood systems into the original INRM framework. The ability of rural people to develop and optimally use their own potential, together with the goal of making a real impact at the farmer level, guided the project's evolution (Hagmann et al. 2003). Once success at the farmer level was evidenced through INRM innovations developed jointly with the farmers, with broader adoption of technical and social innovations, the INRM approach was institutionalized within the extension service. Since 1998, the lessons learned in Zimbabwe have been used to expand and further develop the approach in South Africa. Box 7.4 illustrates the INRM approach with an example of projects in land rehabilitation in the Himalayas.

Box 7.4

Integrated natural resource management (INRM): an example from the Himalayas (Saxena et al. 2003)

The Himalayas are a vast mountain system extending into eight developing countries in South Asia: Afghanistan, Bangladesh, Bhutan, China, India, Myanmar, Nepal, and Pakistan. The fact that India is recognized as a mega-diversity country and as one of the ten most extensively forested areas in the world is due mainly to the Himalayas. Although it covers only 18% of India's geographical area, the Himalayas account for more than 50% of the country's forest cover and 40% of the species endemic to the Indian subcontinent.

Loss of forest cover, biodiversity, agricultural productivity, and ecosystem services in the Himalayan mountain region are interlinked problems and threats to the sustainable livelihoods of 115 million mountain people as well as the inhabitants of the adjoining Indo-gangetic plains. Until the 1970s, environmental conservation, food security, and rural economic development were treated as independent sectors. Development approaches commonly referred to as integrated natural resource research essentially meant the integration of ecological and socioeconomic research, traditional and conventional science, and different actors and stakeholders. However, multiple scales of environmental and development imperatives, from long to short term and from local to global, need to be considered in this type of undertaking. The identification of key natural resource management interventions is an important dimension of integrated management. Although knowledge about the principles and potential advantages of integrated approaches has increased in recent years, there are scientific, technological, and institutional limitations when it comes to putting the theory into practice.

Projects to rehabilitate the degraded lands that cover 40% of the Indian Himalayas could be key interventions provided that they address both socioeconomic and environmental concerns across spatial and temporal scales. However, projects of this type, e.g., investments in conifer plantations on degraded forest lands, have previously failed because their designs did not take into account the needs of local residents, therefore lacking essential elements of the INRM approach. The villagers deliberately damaged the conifer plantations, because they did not want conifers but rather broadleaved species that better served their needs. In addition, legislation in India does not allow the harvesting of conifers on steep hills, and therefore the farmers felt they could not get a direct commercial benefit from these plantations. Conifer plantations also failed for technical reasons: for example, transportation of seedlings from nurseries located in remote regions often damaged the planting material and resulted in poor growth.

This study illustrates a case of land rehabilitation in the small isolated village of Khaljhuni, which is on the margin of the Nanda Devi Biosphere Reserve in the Indian central Himalayas, close to the alpine zone. Vital elements of this project strategy included identifying local perceptions and knowledge and involving the local people in the implementation of the interventions needed to restore the land. Local communities were found to be more concerned with the immediate economic benefit from bamboo and medicinal species than with the long-term benefits of tree planting. The villagers eventually reached a consensus to plant broadleaved multipurpose trees in association with bamboo and medicinal species. Despite assurances that all the economic benefits from rehabilitation would go to the community, the people would not agree to voluntary labor, although they did absorb significant costs by providing social fencing, farmyard manure, and seeds from community forests. Households shared costs and benefits according to traditional norms. The economic benefits to the local people exceeded the rehabilitation cost over the 7-year life of the project. There were significant on-site environmental benefits in terms of improvements in soil fertility, biodiversity, protective cover, and carbon sequestration, and off-site benefits from more productive use of labor, reduced pressure on protected areas, and the introduction of rare and threatened medicinal species to private farmland.

Supplementing indigenous knowledge and the involvement of the whole village community in decision making appeared to be key requirements in the Himalayas for integrating and reconciling diverse concerns about land rehabilitation. The wide range of biophysical and socioeconomic conditions in the Himalayas demands flexible and adaptable approaches to the identification of appropriate rehabilitation technologies.

Experience shows that if INRM is to benefit many people across large areas, considerable political will, investment, and strategic planning from the outset are required. Success depends primarily on building working relationships between technical personnel and local farmers and other members of the participating communities. Community participation has to be ensured, often with the use of some type of incentive (Lovell et al. 2003). INRM aims to identify land-use practices that increase agricultural and forestry production while at the same time maintaining natural capital and continuing to provide environmental services at local and global scales. Once such practices are identified, their adoption by large numbers of people can be facilitated by a combination of educational campaigns and policy changes (Izac and Sanchez 2001).

It could be said that INRM is actually a rediscovery of traditional ways of managing common resources. Freudenberger et al. (1997) describe "tongo", a common property regime that regulates seasonal access to vegetation and wildlife located within village commons and on individually appropriated lands in many areas of The Gambia, Guinea, and Sierra Leone. The system ensures that a particular resource, such as fruits from domesticated and wild trees, or grasses used for thatch reach full maturity before being harvested by the community at large. The authors also give other examples of rules proclaimed by village chiefs restricting hunting, fishing, and cutting grass for thatch to certain seasons. The message is that these community-based rules can be a foundation for working with African indigenous knowledge and institutions to develop an alternative, yet distinctly African approach to resource conservation.

7.7 Importance of Scale in Efficiency of Production

Globalization fosters increase in scale of production units: it is more efficient (by conventional economic standards) to grow 1,000 ha of soybeans than to grow 1 ha, because of economies of scale. Locally centered development is less economically efficient because fields that contain a wide diversity of crops are less amenable to large-scale cultivation. In a globalized economy, big is better than small. However, small can be better than big, when the approach to development is local, as illustrated by the following research projects.

One study carried out in Pernambuco, Brazil, examined two communities established as part of the Movimento dos Trabalhadores Rurais Sem Terra (MST) (Movement of Rural Landless Workers) national movement (March 2001). The objective of this program has been for the people to confiscate agricultural land that is unused or underused, form a community, distribute land among community members, and then demand that government recognizes them under the national settlement programs. The study compared two communities, one that was relatively successful as judged by a survey of members, and another one that was unsuccessful by the same parameters. In the successful community, each family was allotted 2.5 ha, while in the other, the size of the plots was 7.5 ha.

The study showed that the community with the smaller plots was more successful for several reasons. Land-holders in the community with the larger plots were interested more in speculation than in improving agricultural production. They felt that sooner or later, national economic expansion would increase the value of their land, so they devoted most of their effort to ensuring that all of their land was under production. They feared that if any of their land was idle, squatters would move in, or it would be given to someone else. The result was extensive use of the land that stretched management requirements beyond the capabilities of the land-holders. Crops were ill tended. Some farmers hired helpers, but the quality of the work by hired hands was poor. In addition, these farmers used bank credit to buy farm chemicals and seeds, but the bank gave them no choice as to which crops they should plant: they required manioc (*Manihot esculenta*) because the national and international market for this root crop was well established. However, prices for manioc were so low that it was difficult or impossible for the farmers to make enough profit to cover their loans. Communication between community members was also poor. Because of the large area, some members had to walk half a day to attend community meetings, and, consequently, attendance was poor.

In contrast, the farmers with the smaller plots worked their land more intensively, and produced a variety of crops that were more attuned to the continually changing local market. The plots were worked by family members who were more effective than hired hands. Families often incorporated cattle and goats as a source of manure to fertilize their crops and to produce milk and cheese for the local market. Because the plots were smaller, distances were shorter, and there was better communication between members of the community and the community leader.

Another study in the Amazon referred to previously in this chapter showed a similar result (Castellanet and Jordan 2002): in a settlement program for the Amazon, settlers were given 100-ha plots in the region surrounding Altamira. Farms on large areas (100 ha) were less successful than small farms (7 ha). At first, most of the settlers lived on their land, because they wanted to validate their claim to the land. However, because of the poor dirt roads, transportation of agricultural commodities (mostly rice and corn) to the market in Altamira was impossible during the rainy season, when most of the crops were harvested. Also, social services were lacking in the remote regions. Visiting a doctor or going for supplies required days of travel. Access to education was a problem. Consequently, many farmers abandoned their farms, moved closer to Altamira, and bought small parcels of land, typically between 3 and 20 ha. There, with the help of non-governmental organizations, they began planting perennial crops such as black pepper and cocoa that were better suited to the rain forest environment. Shifting cultivation, as was done on the 100-ha plots, was no longer necessary.

7.8 Conclusion

Neither the top-down (authoritarian) nor the bottom-up (participatory) approach alone is capable of solving the problem of how to implement sustainable techniques of forest management and use. What is needed is a mixture of both. Top-down development usually is needed to supply the capital for successful implementation, while the bottom-up approach is necessary to ensure that the project is accepted by the people it was intended to help and that it is environmentally sustainable. Community forestry may be an example of an appropriate blending of the two approaches.

The evidence is mixed with regard to the impact of globalization on global poverty (the income of the poor rises, which is good, but disparity between rich and poor increases, fostering embitterment which is bad) and on the well-being of Third World farmers (mostly benefits big farmers, hurts peasant farmers). However, with regard to the forestry sector, the effects of globalization on tropical forests are almost entirely negative. While there may be some short-term benefit in that industrial logging can temporarily ease a balance of trade problem of a country, after the forest is gone, the social and economic problems are worse. Localization results in a much more rational use of forest products because local knowledge and skills can be brought into play.

Conclusions 8

8.1 Introduction

In the Preface to this book, we asked, are tropical forests more fragile than forests at higher latitudes? Are tropical forests, especially rain forests, particularly susceptible to disturbance? Is recovery following activities such as logging and shifting cultivation slower and more difficult than recovery of temperate zone forests? The answers to these questions have important implications for managing tropical forests for non-timber forest products as well as for wood. To answer these questions, we examined the literature on structure and function of tropical forest ecosystems. Results suggested that tropical forest ecosystems are different from higher latitude forests in ways that make tropical forests more fragile. The five major categories of differences are:

- high diversity of species;
- high frequency of cross-pollination;
- common occurrence of mutualisms;
- high rate of energy flow through primary producers, consumers, and decomposers;
- a relatively tight nutrient cycle.

The high diversity of species means that there are fewer individuals of any species per unit area, and thus the probability of some species becoming locally extinct following logging or other disturbances is relatively high.

The high frequency of cross-pollination means that at least two rather than one individual per species must remain per unit area to prevent losses of genetic diversity, and in some cases even local extinction.

The common occurrence of mutualisms in tropical forest ecosystems means that elimination of one species of plant or animal can result in the elimination of other species of plants or animals. For example, tropical tree species are frequently pollinated by animals that may become extinct due to large-scale disturbances.

The high rate of energy flow through moist tropical ecosystems means that decomposition is rapid, and thus there is high potential for loss of soil organic matter and nutrients. Soil organic matter is critical for the continued health of the whole forest ecosystem.

The tight nutrient cycles on or near the soil surface in moist tropical forests are more easily disrupted by logging and other activities than nutrient cycles in forests where nutrients are held more readily in the deeper mineral soil.

Because of this high susceptibility to disturbance, tropical forests must be managed with special care, if the forests are to reproduce and maintain their productivity in the long term. To ensure reproductive success, care must be taken that sufficient individuals of each species remain in close proximity, and the network of mutualisms must be left relatively intact. To ensure that sufficient organic matter remains on or in the soil, and that nutrient cycling mechanisms remain intact, the basic structure and function of the forest must be maintained.

8.2 Tropical Forest Classification

Not all tropical forests are equally fragile. Forests on nutrient-poor soils are more susceptible to disturbance than those on richer soils, and continually wet forests are sometimes but not always more susceptible than seasonally dry forests. There is no system of classification that satisfactorily predicts the fragility of a forest. Only an understanding of the structure and function of the forest in question can lead to an accurate prediction.

8.3 Tropical Deforestation

The rapid loss of tropical forests is due in part to a lack of understanding of their structure and function, but other factors may be more important, namely economic, political and institutional, technological, cultural, and demographic. In addition, globalization in combination with national debt in developing tropical countries is a major factor driving tropical deforestation.

8.4 Management of Tropical Forests

Systems of natural forest management take advantage of natural ecosystem functions to maintain their productivity and sustain their reproduction. Management techniques include light selective harvest, where the basic structure and function of the forest are left intact. Silvicultural treatments such as girdling unwanted stems may be included, but care must be taken not to destroy the habitat of pollinators and other groups essential for forest functioning. Methods of natural forest management also include cutting of all trees of commercial size, provided that adequate reproduction has been established. Intensive cutting works best with secondary forests that become established in abandoned agricultural areas, or areas of primary forest that have been heavily logged. Improved management techniques such as reduced impact logging and sustainable forest management, in combination with certification of methods of logging, can help slow down the rate of tropical deforestation.

Forest management must also consider activities directed towards the extraction of non-timber forest products (NTFP). Specific guidelines for management for NTFP are needed to ensure that NTFP resources are not overexploited.

Management for conservation of biological diversity is best achieved through minimum disturbance of the forest. If the objective is to save species, the approach should be to conserve and protect forest habitats, thus favoring the successful reproduction and maintenance of plant and animal species.

8.5 Plantations and Agroforestry Systems

Because of their relatively high yields, tropical and subtropical plantations have the potential to make substantial contributions to world timber production. They can also serve other functions in economic development, and environmental services such as carbon sequestration. In addition, plantations, particularly those that establish a mixture of native species, can help maintain the diversity of tropical forests.

Agroforestry systems can contribute to accelerating the return of economic investment and serve social and environmental services as well. Several types of agroforestry systems are practiced worldwide, especially in subsistence economies and for restoration of degraded land.

Strategies that can be used in restoration of degraded primary and secondary forests include enrichment planting; improving soil quality through tree plantations and agroforestry systems; using indigenous species as much as possible; using mixed species plantations to accelerate natural succession; and establishing windbreaks, especially with species that offer habitats for animals that disperse seeds.

8.6 Political and Economic Development Strategies for Sustainable Forest Development

Improved management techniques and plantations that mimic the native forest cannot alone solve the problem of tropical forest loss. For sustainable forest development, there must be capital that frequently comes from development approaches of the top-down type, but capital is not enough. There must be cultural acceptance that can only come from approaches to development of the bottom-up type. The capital necessary to finance successful development is more readily available through the financial connections that arise through globalization.

Until recently, however, and even yet in many cases, the type of development stimulated through globalization has been destructive to both the environment and the local culture. The cultural acceptance necessary for successful implementation is more likely to occur through localized, integrated natural resource management (INRM).

Localized INRM approaches to development offer a feasible alternative. INRM aims to identify land-use practices that increase agricultural and forestry production while at the same time maintaining natural capital and continuing to provide environmental services at local and global scales. Once such practices are identified, their adoption by large numbers of people can be facilitated by a combination of educational campaigns and policy changes.

Ecologically, economically, and socially sustainable development of regions with tropical forests can come about only with a combination of top-down and bottom-up approaches to management. Community forestry may be an example of an appropriate blending of the two approaches. Community forestry projects have been successful in several regions of the world.

In conclusion, tropical forest ecosystems have many characteristics that render them more fragile than forests at other latitudes. At the same time, tropical forests are generally less protected from human disturbance than temperate zone forests, and often are subjected to more intense pressure for development. Therefore, it is important that special care be taken to manage tropical forests in a way that enables their continued existence. In order to ensure sustainability, it is important that forest managers understand the limitations of tropical forest function, and design their management strategies in accordance with these limitations.

References

Aber JD, Melillo JM (1991) Terrestrial ecosystems. Saunders, Philadelphia

Achard F, Eva HD, Stibig HJ, Mayaux P, Gallego J, Richards T, Malingreau JP (2002) Determination of deforestation rates of the world's humid tropical forests. Science 297:999–1002

Acheson JM (1987) The lobster fiefs revisited: economic and ecological effects of territoriality in Maine lobster fishing. In: McCay BJ, Acheson JM (eds) The question of the commons: the culture and ecology of communal resources. University of Arizona Press, Tucson, pp 37–65

Ådjers G, Hadengganan S, Kuusipalo J, Nuryanto K, Vesa L (1995) Enrichment planting of dipterocarps in logged-over secondary forests: effects of width, direction and maintenance method of planting line on selected *Shorea* species. For Ecol Manage 73:259–270

Ae N, Arihara K, Okada T, Yoshihara, Johansen C (1990) Phosphorus uptake by pigeon pea and its role in cropping systems of the Indian subcontinent. Science 248:477–480

Alcorn JB, Molnar A (1996) Deforestation and human–forest relationships: what can we learn from India? In: Sponsel LE, Headland TN, Bailey RC (eds) Tropical deforestation: the human dimension. Columbia University Press, New York, pp 99–121

Allen W (2001) Green Phoenix: restoring the tropical forests of Guanacaste, Costa Rica. Oxford University Press, Oxford

de Almeida JMG (1986) Carajás: desafio político, ecologia e desenvolvimento. Editora Brasiliense SA, São Paulo

Amsallem I, Wilkie ML, Kone P, Ngandji M (eds) (2003) Sustainable management of tropical forests in central Africa. In: Search of excellence. FAO Forestry Pap 143, FAO, Rome, 126 pp

Anderson JM, Proctor J, Vallack HW (1983) Ecological studies in four contrasting lowland rain forests in Bunung Mulu National Park, Sarawakl III: decomposition processes and nutrient losses from leaf litter. J Ecol 71:503–527

Andreisse JP, Schelhaas RM (1987) A monitoring study of nutrient cycles in soils used for shifting cultivation under various climatic conditions in tropical Asia. Agric Ecosyst Environ 19:285–332

Arnold JE (2003) Forests and the people: 25 years of community forestry. FAO, Rome

Ashton PMS (1995) Seedling growth of co-occurring *Shorea* species in the simulated light environments of a rain forest. For Ecol Manage 72:1–12

Ashton PMS, Berlyn GP (1992) Leaf adaptations of some *Shorea* species to sun and shade. New Phytol 121:587–596

Ashton PMS, Gunatilleke CVS, Gunatilleke IAUN (1995) Seedling survival and growth of four *Shorea* species in a Sri Lankan rain forest. J Trop Ecol 11:263–279

Ashton PMS, Gamage S, Gunatilleke IAUN, Gunatilleke CVS (1997a) Restoration of a Sri Lankan rain forest: using Caribbean pine *Pinus caribaea* as a nurse for establishing late-successional species. J Appl Ecol 34:915–925

Ashton PMS, Samarasinghe SJ, Gunatilleke IAUN, Gunatilleke CVS (1997b) Role of legumes in release of successionally arrested grasslands in the central hills of Sri Lanka. Restor Ecol 5:36–43

Ashton PMS, Gamage S, Gunatilleke IAUN, Gunatilleke CVS (1998) Using Caribbean pine to establish a mixed plantation: testing the effects of pine canopy removal on plantings of rain forest tree species. For Ecol Manage 106:211–222

Ashton PMS, Gunatilleke CVS, Singhakumara BMP, Gunatilleke IAUN (2001) Restoration pathways for rain forest in southwest Sri Lanka: a review of concepts and models. For Ecol Manage 525:1–23

Ashton PS (2000) Ecological theory of diversity and its application to mixed species plantation systems. In: Ashton MS, Montagnini F (eds) The silvicultural basis for agroforestry systems. CRC Press, Boca Raton, pp 61–77

Ashton PS, Ducey MJ (2000) Agroforestry systems as successional analogs to native forests. In: Ashton MS, Montagnini F (eds) The silvicultural basis for agroforestry systems. CRC Press, Boca Raton, pp 207–228

Ba TA, Crousse B (1985) Food-production systems in the Middle Valley of the Senegal River. Int Social Sci J 37:389–400

Bailey RC (1996) Promoting biodiversity and empowering local peoples in Central African forests. In: Sponsel NE, Headland TE, Bailey RC (eds) Tropical deforestation: the human dimension. Columbia University Press, New York, pp 316–341

Bais HP, Vepachedu R, Gilroy S, Callaway RM, Vivanco JM (2003) Allelopathy and exotic plant invasion: from molecules and genes to species interactions. Science 301:1377–1380

Barbosa LC (2000) The Brazilian Amazon rainforest. University Press of America, Lanham, Maryland

Barquero M (2004) Fuerte caída en la reforestación anual. La Nación, Costa Rica, 19 Jan (www.nacion.com)

Bates HW (1864) Animals of Ega. In: Murray J (ed) The Naturalist on the River Amazonas. John Murray, London, pp 412–426

Batmanian G (1990) Reforestation of degraded pastures in the Brazilian Amazon: effect of site preparation on phosphorus availability in the soil. PhD Diss, University of Georgia, Athens, Georgia

Bawa KS (1992) Mating systems, genetic differentiation and speciation in tropical rain forest plants. Biotropica 24:250–255

Bawa KS, Krugman SL (1991) Reproductive biology and genetics of tropical trees in relation to conservation and management. In: Gómez-Pompa A, Whitmore TC, Hadley M (eds) Rainforest regeneration and management. Man and the Biosphere series, vol 6. UNESCO, Paris, and the Parthenon Publishing Group, Carnforth, UK, pp 119–136

Bayliss-Smith T, Hviding E, Whitmore T (2003) Rainforest composition and histories of human disturbance in the Solomon Islands. Ambio 32:346–352

Beard JS (1949) The natural vegetation of the Windward and Leeward Islands. Clarendon Press, Oxford

Beer J (1988) Litter production and nutrient cycling in coffee (*Coffea arabica*) or cacao (*Theobroma cacao*) plantations with shade trees. Agrofor Syst 7:33–45

Beer J, Harvey CA, Ibrahim M, Harmand JM, Somarriba E, Jimenez J (2003) Service functions of agroforestry systems. In: Proc 12th World Forestry Congr, Area B: Forests for the Planet, Quebec, Canada, 21–28 Sept, pp 417–424

Bennett EI (2000) Timber certification: where is the voice of the biologist? Conserv Biol 14:921–923
Bertault JG, Sist P (1997) An experimental comparison of different harvesting intensities with reduced-impact and conventional logging in East Kalimantan, Indonesia. For Ecol Manage 94:209–218
Binkley D, Dunkin KA, De Bell D, Ryan MG (1992) Production and nutrient cycling in mixed plantations of *Eucalyptus* and *Albizia* in Hawaii. For Sci 38:393–408
Bishop J, Landell-Mills N (2002) Forest environmental services: an overview. In: Pagiola S, Bishop J, Landell-Mills N (eds) Selling forest environmental services: market-based mechanisms for conservation and development. Earthscan, London, pp 15–35
Boucher DH (1985) Mutualism in agriculture. In: Boucher DH (ed) The biology of mutualism. Oxford Univ Press, New York, pp 375–386
Bowles IE, Rice RE, Mittermeier RA, da Fonseca GAB (1998) Logging and tropical forest conservation. Science 280:1899–1900
Boyle TJB, Sayer JA (1995) Measuring, monitoring and conserving biodiversity in managed tropical forests. Comm For Rev 74:20–25
Boza MA (2001) Corredores ecológicos de las Américas: una vía verde para conservar la naturaleza y contribuir con el desarrollo de los estados americanos. Biocenosis 15:40–48
Brady NC, Weil RR (2002) The nature and properties of soils, 13th edn. Prentice Hall, Englewood Cliffs
Brandon K, Redford K, Sanderson S (eds) (1998) Parks in peril: people, politics and protected areas. Island Press, Covelo
Bray DB (2003) Mexico's community-managed forests as a global model for sustainable landscapes. Conserv Biol 17:672–677
Bronstein JL, Wilson WG, Morris WF (2003) Ecological dynamics of mutualist/antagonist communities. Am Nat 162(Suppl):S24–S39
Brookfield HC (1993) Farming the forests in Southeast Asia. In: Proc UNU Global Environmental Forum II, Environmental Change in Rain Forests and Drylands, United Nations University, Tokyo, pp 51–70
Brosius JP (1995) Trivializing indigenous resistance to deforestation: Sarawak Malaysia. In: Jordan CF (ed) Conservation: replacing quantity with quality as a goal for global management. Wiley, New York, pp 260–261
Brown N (1998) Out of control: fires and forestry in Indonesia. Trends Ecol Evol 13:41
Brown S, Iverson LR (1992) Biomass estimates for tropical forests. World Resour Rev 4:366–383
Brown S, Lugo AE (1982) The storage and production of organic matter in tropical forests and their role in the global carbon cycle. Biotropica 14:161–187
Brown S, Lugo AE (1990) Tropical secondary forests. J Trop Ecol 6:1–32
Brown S, Lugo AE (1994) Rehabilitation of tropical lands: a key to sustaining development. Restor Ecol 2:97–111
Brown S, Gillespie AJR, Lugo AE (1989) Biomass estimation methods for tropical forests with applications to forest inventory data. For Sci 35:881–902
Bruenig EF (1996) Conservation and management of tropical rainforests: an integrated approach to sustainability. CABI International, Wallingford
Bruijnzeel LA, Veneklaas EJ (1998) Climatic conditions and tropical montane forest productivity: the fog has not lifted yet. Ecology 79:3–9
Bruna EM (1999) Seed germination in rainforest fragments. Nature 402:139

Budowski G (1965) Distribution of tropical rainforest species in the light of successional process. Turrialba 15:40–42
Burkhart HE, Tham Å (1992) Predictions from growth and yield models of the performance of mixed-species stands. In: Cannell MGR, Malcolm DC, Robertson PA (eds) The ecology of mixed-species stands of trees. Blackwell, Boston, pp 21–34
Buschbacher RJ (1986) Tropical deforestation and pasture development. BioScience 36:22–28
Buschbacher RJ (1990) Natural forest management in the humid tropics: ecological, social, and economic considerations. Ambio 19:253–258
Byard R, Lewis LC, Montagnini F (1996) Leaf litter decomposition and mulch performance from mixed and monospecific plantations of native tree species in Costa Rica. Agric Ecosyst Environ 58:145–155
Cairns MA, Meganck RA (1994) Carbon sequestration, biological diversity, and sustainable development: integrated forest management. Environ Manage 18:13–22
de Camino T (1997) Utilización de sistemas de información para muestreo, procesamiento y análisis de información. In: Resumen de ponencias III Congreso Forestal de Costa Rica, 27–29 Aug, San José, Costa Rica
Campos JJ, Ortíz R (1999) Capacidad y riesgos de actividades forestales en el almacenamiento de carbono y conservación de biodiversidad en fincas privadas del área central de Costa Rica. In: Proc 4th Semana Científica, Logros de la Investigación para el Nuevo Milenio, 6–9 April, Programa de Investigación, Centro Agronómico Tropical de Investigación y Enseñanza (CATIE), Turrialba, Costa Rica, pp 291–294
Campos JJ, Finegan B, Camacho M, Quirós D (1998) Sostenibilidad del manejo de bosques naturales: resultados sobre la factibilidad ecológica y económica en Costa Rica. In: Trabajo presentado al Primer Congreso Latinoamericano IUFRO, Valdivia, Chile
Carle J, Vuorinen P, Del Lungo A (2002) Status and trends in global forest plantation development. For Products J 52:12–23
Carneiro RL (1988) Indians of the Amazonian forest. In: Denslow JS, Padoch C (eds) People of the tropical rain forest. University of California Press, Berkeley, pp 73–86
Carnevale NJ, Montagnini F (2002) Facilitating regeneration of secondary forests with the use of mixed and pure plantations of indigenous tree species. For Ecol Manage 163:217–227
Carnus J-M, Parrotta J, Brockerhoff EG, Arbez M, Jactel H, Kremer A, Lamb D, O'Hara K, Walters B (2003) Planted forests and biodiversity. In: Buck A, Parrotta J, Eolfrum G (eds) Science and technology – building the future of the world's forests: planted forests and biodiversity. IUFRO Occasional Pap 15. IUFRO, Vienna, pp 33–49
Castellanet C, Jordan CF (2002) Participatory action research in natural resource management. Taylor and Francis, New York
Cattaneo A (2001) A general equilibrium analysis of technology, migration, and deforestation in the Brazilian Amazon. In: Angelsen A, Kaimowitz D (eds) Agricultural technologies and tropical deforestation. CABI, Wallingford, pp 69–90
Center for Resources and Environmental Studies (2003) Improving smallholder farming systems in *Imperata* areas of Southeast Asia – a bioeconomic modeling approach. Institute of Advanced Studies, Australian National University, http://incres.anu.edu.au/imperata/imperat1. htm
CGIAR-INRM Group (1999) Integrated natural resource management: the Bilderberg consensus. In: Summary Rep of INRM Worksh, Bilderberg, The Netherlands, 3–5 Sept, Consultative Group for International Agricultural Research-Integrated Natural Resource Management, http://www.inrm.cgiar.org/

Chambers R, Leach M (1990) Trees as savings and security for the rural poor. Unasylva 41:39–52
Chambers R, Pacey RA, Thrupp LA (eds) (1989) Farmers first: farmer innovation and agricultural research. Intermediate Technology Publications, London
Chapin FS (1980) The mineral nutrition of wild plants. Annu Rev Ecol Syst 11:233–260
Chapman CA, Chapman LJ (1996) Exotic tree plantations and the regeneration of natural forests in Kibale National Park, Uganda. Biol Conserv 76:253–257
Charley JL, Richards BN (1983) Nutrient allocation in plant communities: mineral cycling in terrestrial ecosystems. In: Lange OL, Nobel PS, Osmond CB, Ziegler H (eds) Physiological plant ecology, vol 4. Ecosystem processes: mineral cycling, productivity, and man's influence. Springer, Berlin Heidelberg New York, pp 5–45
Chazdon RL, Earl of Cranbrook (2002) Tropical naturalists of the sixteenth through nineteenth centuries. In: Chazdon RL, Whitmore TC (eds) Foundations of tropical forest biology. University of Chicago Press, Chicago, pp 5–14
Chazdon RL, Fetcher N (1984) Light environments of tropical forests. In: Medina E, Mooney HA, Vázquez-Yanes C (eds) Physiological ecology of plants of the wet tropics. W Junk, The Hague, pp 553–564
Chazdon RL, Pearcy RW (1991) The importance of sunflecks for forest understory plants. BioScience 41:760–766
Chazdon RL, Pearcy RW, Lee DW, Fetcher N (1996) Photosynthetic responses of tropical forest plants to contrasting light environments. In: Mulkey SS, Chazdon RL, Smith AP (eds) Tropical forest plant ecophysiology. Chapman and Hall, New York, pp 5–55
Chokkalingam U, Smith J, de Jong W (2001) A conceptual framework for the assessment of tropical secondary forest dynamics and sustainable development potential in Asia. J Trop For Sci 13:577–600
Chomitz K, Kumari K (1995) The domestic benefits of tropical forests: a critical review emphasizing hydrological functions. Policy Research Working Pap 1601. Policy Research Department, World Bank, Washington, DC
Clark HL, Liesner RL (1989) Angiosperms. In: Jordan CF (ed) An Amazon rain forest: the structure and function of a nutrient-stressed ecosystem and the impact of slash-and-burn agriculture. Man and the Biosphere Series, vol 2. UNESCO, Paris, and The Parthenon Publishing Group, Carnforth, UK, pp 111–120
Cobbina J (1994/1995) Growth and herbage productivity of *Gliricidia* in *Panicum maximum* pasture as influenced by seed preparation, planting and weed control techniques. Agrofor Syst 28:193–201
Cohen AL, Singhakumara BMP, Ashton PMS (1995) Releasing rain forest succession: a case study in the *Dicranopteris linearis* fernlands of Sri Lanka. Restor Ecol 3:261–270
Coley PD (1980) Effects of leaf age and plant life history patterns on herbivory. Nature 284:545–546
Coley PD (1982) Rates of herbivory on different tropical trees. In: Leigh EG, Rand AS, Windson DM (eds) The ecology of a tropical forest. Smithsonian Institution Press, Washington, DC, pp 123–132
Coley PD (1983) Herbivory and defensive characteristics of tree species in a lowland tropical forest. Ecol Monogr 53:209–233
Coley PD, Barone JA (1996) Herbivory and plant defenses in tropical forests. Annu Rev Ecol Syst 27:305–335
Condit R, Pitman N, Leigh EG, Chave J, Terborgh J, Foster RB, Nunez P, Aguilar S, Nalencia R, Villa G, Muller-Landau HC, Losos E, Hubbell SP (2002) Beta-diversity in tropical forest trees. Science 295:666–669

Connell JH (1978) Diversity in tropical forests and coral reefs. Science 199:1302–1310
Connell JH, Orias E (1964) The ecological regulation of species diversity. Am Nat 98:399–414
de Cordoba J (2004) Along the Andes, Indians agitate for political gain. Wall Street Journal, 8 Jan, p 1
Cossalter C, Pye-Smith C (2003) Fast-wood forestry: myths and realities. Forest perspectives. Center for International Forestry Research (CIFOR), Jakarta, Indonesia, 50 pp
Coxhead I, Shively G, Shuai X (2001) Agricultural development policies and land expansion in a southern Philippine watershed. In: Angelsen A, Kaimowitz D (eds) Agricultural technologies and tropical deforestation. CABI, Wallingford, pp 347–365
Cuellar ER (2002) Concesiones forestales comunitarias en la Reserva de la Biosfera Maya, Peten, Guatemala. In: Actas del II Congreso Forestal Latinoamericano. Bienes y servicios del bosque, fuente de desarrollo sostenible. Guatemala, 31 July–2 Aug, CD-ROM
Cuevas E (2001) Soil versus biological controls on nutrient cycling in terra firme forests. In: McClain ME, Victoria RL, Richey JE (eds) Biogeochemistry of the Amazon Basin. Oxford University Press, Oxford, pp 53–67
Curran LM, Trigg SN, McDonald AK, Astiani D, Hardiono YM, Siregar P, Caniago I, Kasischke E (2004) Lowland forest loss in protected areas of Indonesian Borneo. Science 303:1000–1003
Current D, Rossi LMB, Sabogal C, Nalvarte W (1998) Comparación del potencial del manejo de la regeneración natural con asocio agroforestal y plantaciones puras para tres especies: estudios de caso en Brasil, Perú y Costa Rica. In: Trabajo presentado al Primer Congreso Latinoamericano IUFRO, Valdivia, Chile
Cusack D, Montagnini F (2004) The role of native species plantations in recovery of understory diversity in degraded pasturelands of Costa Rica. For Ecol Manage 188:1–15
Cusack D, Hodgdon BD, Montagnini F (eds) (2002) Forests, communities and sustainable management. A summary of a forum examining community forestry initiatives in the tropics. YFF Review 5(6). GISF, Yale University, School of Forestry and Environmental Studies, New Haven
Dabas M, Bhatia S (1996) Carbon sequestration through afforestation: role of tropical industrial plantations. Ambio 25:327–330
Daly HE (1996) Free trade: the perils of deregulation. In: Mander J, Goldsmith E (eds) The case against the global economy. Sierra Club Books, San Francisco, pp 229–238
Daly HE (2001) Globalization and its discontents. Philos Public Policy Q 21:17–21
Darwin C (1855) Journal of researches into the natural history and geology of the countries visited during the voyage of H.M.S. Beagle round the world. Appleton, New York, 389 pp
Davis TA, Richards PW (1933/1934) The vegetation of Moraballi Creek, British Guiana: an ecological study of a limited area of tropical rain forest, parts I and II. J Ecol 21:350–384; 22:106–155
Dawkins HC (1961) New methods of improving stand composition in tropical forests. Carib For 22:12–20
Dawkins HC, Philip MS (1998) Tropical moist forest silviculture and management: a history of success and failure. CAB International, Wallingford
Deevey ES (1949) Biogeograpy of the Pleistocene. Bull Geol Soc Am 60:1315–1416

Dekker M, De Graaf NR (2003) Pioneer and climax tree regeneration following selective logging with silviculture in Surinam. For Ecol Manage 172:183–190
Denslow JS, Padoch C (1988) (eds) People of the tropical rain forest. University of California Press, Berkeley, 232 pp
De Onis MC, Monteiro C, Akré J, Clugston G (2001) The worldwide magnitude of protein-energy malnutrition: an overview from the WHO Global Database on Child Growth (1 May 2002; www/who.int/whosis/cgrowth/bulletin.htm)
Dirzo R, Raven PH (2003) Global state of biodiversity and loss. Annu Rev Environ Resour 28:02.1–02.31
Dixon RK (1995) Agroforestry systems: sources or sinks of greenhouse gases? Agrofor Syst 31:99–116
Dobzhansky T (1950) Evolution in the tropics. Am Sci 38:209–221
Dolanc CR, Gorchov DL, Cornejo F (2003) The effects of silvicultural thinning on trees regenerating in strip clear-cuts in the Peruvian Amazon. For Ecol Manage 182:103–116
Dorman P (2003) Debt and deforestation. In: Harris JM, Goodwin NR (eds) New thinking in macroeconomics. Edward Elgar, Cheltenham, UK, pp 213–228
Dubois JCL (1990) Secondary forests as a land-use resource in frontier zones of Amazonia. In: Anderson AB (ed) Alternatives to deforestation: steps toward sustainable use of the Amazon rain forest. Columbia University Press, New York, pp 183–194
Dudley NS, Fownes JH (1992) Preliminary biomass equations for eight species of fast-growing tropical trees. J Trop For Sci 5:68–73
Duivenvoorden JF, Svenning JC, Wright SJ (2002) Beta diversity in tropical forests. Science 295:636–637
The Economist (2000) The case for globalization. The Economist, 23 Sept, pp 19–20
Eibl B, Fernández R, Kozarik JC, Lupi A, Montagnini F, Nozzi D (2000) Agroforestry systems with *Ilex paraguariensis* (American holly or yerba mate) and native timber trees on small farms in Misiones, Argentina. Agrofor Syst 48:1–8
Erwin TL (1982) Tropical forests: their richness in Coleoptera and other arthropod species. Coleopt Bull 36:74–75
Erwin TL (1988) The tropical forest canopy: the heart of biotic diversity. In: Wilson EO, Peters FM (eds) Biodiversity. National Academy Press, Washington, DC, pp 123–129
Evans J (1992) Plantation forestry in the tropics. Clarendon Press, Oxford
Evans J (1999) Planted forests of the wet and dry tropics: their variety, nature, and significance. New For 17:25–36
Fals-Borda O, Rahman MA (1991) Action and knowledge: breaking the monopoly with participatory action-research. Apex Press, New York
FAO (1989) Production yearbook 1987. FAO, Rome
FAO (2000 a) Statistical databases. Roundwood, sawnwood, wood-based panels. http://apps.fao.org
FAO (2000 b) Global forest resources assessment 2000. Main report. http:/www/fao.org/forestry/fo/fra/main
FAO (2001) State of the world's forests 2001. FAO, Rome
Farnworth EG, Golley FB (1974) Fragile ecosystems: evaluation of research and applications in the Neotropics. Springer, Berlin Heidelberg New York
Farnworth EG, Tidrick TH, Jordan CF, Smathers WM (1981) The value of natural ecosystems: an economic and ecological framework. Environ Conserv 8:275–281
Fay JM (1997) Development of a trinational system of conservation: a ten-year perspective. In: Proc Sangha River Network Conf, Yale University 25–28 Sept, Yale For Environ Stud Bull 102:253–258

Fearnside PM (2003) Conservation policy in Brazilian Amazonia: understanding the dilemmas. World Dev 31:757–779
Fearnside PM, Laurance WF (2003) Comment on "Determination of deforestation rates of the world's humid tropical forests", http://www.sciencemag.org/cgi/content/full/299/5609/1015a
Fernandes ECM, O'Kting'ati A, Maghembe J (1989) The Chagga homegardens: a multistoreyed agroforestry cropping system on Mount Kilimanjaro (northern Tanzania). In: Nair PKR (ed) Agroforestry systems in the tropics. Kluwer, Dordrecht, pp 309–332
Fernández R, Montagnini F, Hamilton H (1997) The influence of native tree species on soil chemistry in a subtropical humid forest region of Argentina. J Trop For Sci 10:188–196
Fimbel RA, Fimbel CC (1996) The role of exotic conifer plantations in rehabilitating degraded tropical forest lands: a case study from the Kibale Forest in Uganda. For Ecol Manage 81:215–226
Finegan B (1992) The management potential of neo-tropical secondary lowland rainforest. For Ecol Manage 47:295–321
Fischer AG (1960) Latitudinal variations in organic diversity. Evolution 14:64–81
Flores EM (1993) Arboles y semillas del Neotropico, vol 2(1). Museo Nacional de Costa Rica, Herbario Nacional, San José, Costa Rica
Fölster H, Khanna PK (1997) Dynamics of nutrient supply in plantation soils. In: Nambiar EKS, Brown AG (eds) Management of soil, nutrients and water in tropical plantation forests. ACIAR/CSIRO/CIFOR. ACIAR, Canberra, Australia, pp 338–378
Forman RT (1975) Canopy lichens with blue-green algae: a nitrogen source in a Colombian rain forest. Ecology 56:1176–1184
Fox JED (1976) Constraints on the natural regeneration of tropical moist forest. For Ecol Manage 1:37–65
Frederiksen TS, Putz FE (2003) Silvicultural intensification for tropical forest conservation. Biodiversity Conserv 12:1445–1453
Freire P (1970) Pedagogy of the oppressed. Harper and Herder, New York
Freudenberger MS, Carney JA, Lebbie AR (1997) Resiliency and change in common property regimes in West Africa: the case of the Tongo in The Gambia, Guinea, and Sierra Leone. Soc Nat Resour 10:383–402
Friedman TL (2004) Meet the Zippies. The New York Times, 22 Feb, Sect 4 (Week in Review), p 11
Gajaseni J (1992) Overview of Taungya. In: Jordan CF, Gajaseni J, Watanabe H (eds) Taungya: forest plantations with agriculture in Southeast Asia. CAB International, Wallingford, pp 3–8
Gajaseni J, Jordan CF (1992) Theoretical basis for Taungya and its improvement. In: Jordan CF, Gajaseni J, Watanabe H (eds) Taungya: forest plantations with agriculture in Southeast Asia. CAB International, Wallingford, pp 68–81
Geist HJ, Lambin EF (2002) Proximate causes and underlying driving forces of tropical deforestation. BioScience 52:143–150
Gentry AH (1988) Changes in plant community diversity and floristic composition on environmental and geographical gradients. Ann Mo Bot Gard 75:1–3
Gilbert LE (1980) Food web organization and the conservation of neotropical diversity. In: Soulé ME, Wilcox BZ (eds) Conservation biology: an evolutionary–ecological perspective. Sinauer Associates, Sunderland, Massachusetts, pp 11–33
Gill AS, Gangwar KS, Sinsinwar BS (1990) Productivity of perennial grasses in association with *Acacia albida* under different cutting schedules in dryland conditions. Indian J Dryland Agric Res Dev 5:68–71

Gillis M, Repetto R (1988) Conclusion: findings and policy implications. In: Repetto R, Gillis M (eds) Public policies and the misuse of forest resources. Cambridge University Press, Cambridge, pp 385–410
Gobierno de la Provincia de Misiones, Argentina (2003 a) Estadísticas forestales. Subsecretaría de Bosques y Forestaciones, Ministerio de Ecología, Recursos Naturales y Turismo, Posadas, Misiones
Gobierno de la Provincia de Misiones, Argentina (2003 b) Actitud Forestal. Un mundo de oportunidades para una Misiones estratégica. Subsecretaría de Bosques y Forestaciones, Ministerio de Ecología, Recursos Naturales y Turismo, Posadas, Misiones
Goldhammer JG (1993) Fire management. In: Pancel L (ed) Tropical forestry handbook. Springer, Berlin Heidelberg New York, pp 1221–1268
Goldsmith E (1996 a) Development as colonialism. In: Mander J, Goldsmith E (eds) The case against the global economy. Sierra Club Books, San Francisco, pp 253–266
Goldsmith E (1996 b) Global trade and the environment. In: Mander J, Goldsmith E (eds) The case against the global economy. Sierra Club Books, San Francisco, pp 78–91
Golley F (1977) Insects as regulators of forest nutrient cycling. Trop Ecol 1:116–124
Golley FG, McGinnis JT, Clements RG, Child GI, Deuver MJ (1975) Mineral cycling in a tropical moist forest ecosystem. University of Georgia Press, Athens, Georgia
Gómez-Pompa A (1991) Learning from traditional ecological knowledge: insights from Mayan silviculture. In: Gómez-Pompa A, Whitmore TC, Hadley M (eds) Rainforest regeneration and management. MAB Series, UNESCO, Paris, pp 335–342
Gómez-Pompa A, Vázquez-Yanes C, Guevara S (1972) The tropical rain forest: a non-renewable resource. Science 177:762–765
de Graaf NR (2000) Reduced impact logging as part of domestication of neotropical rain forest. Int For Rev 2:40–44
De Graaf NR, Poels RLH (1990) The Celos management system: a polycyclic method for sustained timber production in South American rain forest. In: Anderson AB (ed) Alternatives to deforestation: steps toward sustainable use of the Amazon rain forest. Columbia University Press, New York, pp 116–127
Grainger A (1993) Controlling tropical deforestation. Earthscan, London
Grubb PJ (1977) Control of forest growth and distribution on wet tropical mountains, with special reference to mineral nutrition. Annu Rev Ecol Syst 8:83–107
Grubb PJ (1995) Mineral nutrients and soil fertility in tropical rain forests. In: Lugo AE, Lowe C (eds) Tropical forests: management and ecology. Ecological Studies 112. Springer, Berlin Heidelberg New York, pp 308–330
Guariguata MR (1999) Early response of selected tree species to liberation thinning in a young secondary forest in northeastern Costa Rica. For Ecol Manage 127:255–261
Guariguata MR (2000) Seed and seedling ecology of tree species in neotropical secondary forests: management implications. Ecol Appl 10:145–154
Guariguata MR, Rheingans R, Montagnini F (1995) Early woody invasion under tree plantations in Costa Rica: implications for forest restoration. Restor Ecol 3:252–260
Guariguata MR, Rosales JJ, Finegan B (2000) Seed removal and fate in two selectively-logged lowland forests with contrasting protection levels. Conserv Biol 14:1046–1054
Guindon C (1996) The importance of forest fragments to the maintenance of regional biodiversity in Costa Rica. In: Schelhas J, Greenberg R (eds) Forest patches in tropical landscapes. Island Press, Washington, DC, pp 168–186
Hagmann J, Chuma E, Murwira K, Connolly M, Ficarelli PP (2003) Success factors in integrated natural resource management R & D: lessons from practice. In: Camp-

bell BM, Sayer JA (eds) Integrated natural resource management. Linking productivity, the environment and development. CABI, Wallingford, pp 37–64
Hall AL (1989) Developing Amazonia: deforestation and social conflict in Brazil's Carajás programme. Manchester University Press, Manchester
Hall JS, Harris DJ, Medjibe V, Ashton M (2003) The effects of selective logging on forest structure and tree species composition in a central African forest: implications for management and conservation areas. For Ecol Manage 183:249–264
Halstead P, O'Shea J (1989) Introduction: cultural responses to risk and uncertainty. In: Halstead P, O'Shea J (eds) Bad year economics: cultural responses to risk and uncertainty. Cambridge University Press, New York, pp 1–7
Hardin G (1968) The tragedy of the commons. Science 162:1243–1248
Hartshorn GS (1990) Natural forest management by the Yanesha forestry cooperative in Peruvian Amazonia. In: Anderson AB (ed) Alternatives to deforestation: steps toward sustainable use of the Amazon rain forest. Columbia University Press, New York, pp 128–138
Hartshorn GS, Hammel BE (1994) Vegetation types and floristic patterns. In: McDade LA, Bawa K, Hespenheide HA, Hartshorn GS (eds) La Selva: ecology and natural history of a neotropical rainforest. University of Chicago Press, Chicago, pp 73–89
Harvey CA (2000) Windbreaks enhance seed dispersal into agricultural landscapes in Monteverde, Costa Rica. Ecol Appl 10:155–173
Harvey CA, Haber WH (1999) Remnant trees and the conservation of biodiversity in Costa Rican pastures. Agrofor Syst 44:37–68
Hecht SB (1982) Cattle ranching development in the eastern Amazon: evaluation of a development strategy. PhD Diss, Department of Geography, University of California at Los Angeles
Hecht SB (1984) Cattle ranching in Amazonia: political and ecological considerations. In: Schmink M, Wood CH (eds) Frontier expansion in Amazonia. University of Florida Press, Gainesville, pp 366–398
Hecht SB (1985) Environment, development and politics: capital accumulation and the livestock sector in eastern Amazonia. World Dev 13(6):663–684
Heckenberger MJ, Kuikuro A, Tabata Kuikuro U, Russell JC, Schmidt M, Fausto C, Franchetto B (2003) Amazonia 1492: pristine forest or cultural parkland? Science 301:1710–1714
Herrera R, Merida T, Stark N, Jordan C (1978) Direct phosphorus transfer from leaf litter to roots. Naturwissenschaften 65:208–209
Higman S, Bass S, Judd N, Mayers J, Nussbaum R (1999) The sustainable forestry handbook. Earthscan, London
Holdridge LR (1967) Life zone ecology. Revised edition. Tropical Science Center, San José, Costa Rica
Holl KD (1998) Do bird perching structures elevate seed rain and seedling establishment in abandoned tropical pasture? Restor Ecol 6:253–261
Holl KD (1999) Factors limiting tropical rain forest regeneration in abandoned pasture: seed rain, seed germination, microclimate and soil. Biotropica 31:229–242
Holmes TP, Blate GM, Zweede JC, Pereira R Jr, Barreto P, Boltz F, Bauch R (2002) Financial and ecological indicators of reduced impact logging performance in the eastern Amazon. For Ecol Manage 163:93–110
Hosington D, Khairallah M, Reeves T, Ribaut J-M, Skovmand B, Taba S, Warburton M (1999) Plant genetic resources: what can they contribute toward increased crop productivity? Proc Natl Acad Sci 96:5937–5943

Hubbell SP (2001) The unified neutral theory of biodiversity and biogeography. Princeton University Press, Princeton
Huston M (1979) A general hypothesis of species diversity. Am Nat 113:81–101
International Tropical Timber Organization (2002) Guidelines for the restoration, management and rehabilitation of degraded and secondary tropical forest. ITTO Policy Development Ser 13. ITTO, Yokohama, 84 pp
International Tropical Timber Organization (2003a) Assessing progress towards sustainable forest management in the tropics. www.itto.or.jp/inside/measuring_up/index.html
International Tropical Timber Organization (2003b) Criteria and indicators for sustainable management of natural tropical forests. www.itto.or.jp/policy/pds7
Intergovernmental Panel on Climatic Change (2000) Special report on land use, land use change and forestry. Summary for policy makers. IPCC, Geneva, 20 pp
Irwin DA (2004) "Outsourcing" is good for America. Wall Street Journal, 28 Jan, p A16
Izac AMN, Sanchez PA (2001) Towards a natural resource management paradigm for international agriculture: the example of agroforestry research. Agric Syst 69:5–25
Jackson ET, Kassam Y (1998) Knowledge shared: participatory evaluation in development cooperation. Kumarian Press, West Hartford, Connecticut
Jackson RM, Raw F (1973) Life in the soil. Edward Arnold, London
Jacobs M (1988) The tropical rainforest: a first encounter. Springer, Berlin Heidelberg New York
Janzen DH (1967) Why mountain passes are higher in the tropics. Am Nat 101:233–249
Janzen DH (1970) Herbivores and the number of tree species in tropical forests. Am Nat 104:501–528
Janzen DH (1974) Tropical blackwater rivers, animals, and mast fruiting by the Dipterocarpaceae. Biotropica 6:69–103
Janzen DH (1985) The natural history of mutualisms. In: Boucher DH (ed) The biology of mutualism. Oxford University Press, New York, pp 40–99
Janzen DH, Vázquez-Yanes C (1991) Aspects of tropical seed ecology of relevance to management of tropical forested wildlands. In: Gómez-Pompa A, Whitmore TC, Hadley M (eds) Rain forest regeneration and management. Man and the Biosphere Series, vol 6. UNESCO, Paris, and The Parthenon Publishing Group, Carnforth, UK, pp 137–157
Johnson DW, Cole DW, Gessel SP, Singer MJ, Minden MV (1977) Carbonic acid leaching in a tropical, temperate, subalpine, and northern forest soil. Arctic Alpine Res 9:329–343
Jordan CF (1971a) A world pattern in plant energetics. Am Sci 59:425–433
Jordan CF (1971b) Productivity of a tropical forest and its relation to a world pattern of energy storage. J Ecol 59:127–142
Jordan CF (1982) Amazon rain forests. Am Sci 70:394–701
Jordan CF (1983) Productivity of tropical rain forest ecosystems and the implications for their use as future wood and energy sources. In: Golley FB (ed) Tropical rain forest ecosystems. Ecosystems of the world, vol 14A. Elsevier, Amsterdam
Jordan CF (1985) Nutrient cycling in tropical forest ecosystems. Wiley, Chichester
Jordan CF (1989) An Amazon rain forest: the structure and function of a nutrient stressed ecosystem and the impact of slash-and-burn agriculture. Man and the Biosphere Series, vol 2. UNESCO, Paris, and The Parthenon Publishing Group, Carnforth, UK

Jordan CF (1993) Ecology of tropical forests. In: Pancel L (ed) Tropical forestry handbook, vol 1. Springer, Berlin Heidelberg New York, pp 164–197
Jordan CF (1995a) Nutrient cycling in tropical forests. In: Encyclopedia of environmental biology. Academic Press, New York, pp 641–654
Jordan CF (1995b) Conservation: replacing quantity with quality as a goal for global management. Wiley, New York
Jordan CF (1998) Working with nature. Harwood Academic, Amsterdam
Jordan CF (2001) The interface between economics and nutrient cycling in Amazon land development. In: McClain ME, Victoria RL, Richey JE (eds) The Biogeochemistry of the Amazon Basin. Oxford University Press, New York, pp 156–164
Jordan CF (2002) Genetic engineering, the farm crisis, and world hunger. BioScience 52:523–529
Jordan CF, Farnworth EG (1980) A rain forest chronicle: perpetuation of a myth. Biotropica 12:233–234
Jordan CF, Farnworth EG (1982) Natural vs plantation forests: a case study of land reclamation strategies for the humid tropics. Environ Manage 6:485–492
Jordan CF, Murphy PG (1978) A latitudinal gradient of wood and litter production, and its implication regarding competition and species diversity in trees. Am Midl Nat 99:415–434
Jordan CF, Golley F, Hall JD, Hall J (1979) Nutrient scavenging of rainfall by the canopy of an Amazonian rain forest. Biotropica 12:61–66
Jordan CF, Caskey W, Escalante G, Herrera R, Montagnini F, Todd R, Uhl C (1982) The nitrogen cycle in a "tierra firme" rainforest on oxisol in the Amazon Territory of Venezuela. Plant Soil 67:325–332
Jordan CF, Gajaseni J, Watanabe H (eds) (1992) Taungya: forest plantations with agriculture in Southeast Asia. CAB International. Wallingford, 152 pp
Kahn JR, McDonald JA (1995) Third-World debt and tropical deforestation. Ecol Econ 12:107–123
Kaimowitz D (1996) Livestock and deforestation. Central America in the 1980s and 1990s: a policy perspective. CIFOR Spec Publ. Center for International Forestry Research, Bogor, Indonesia
Kammesheidt L (2002) Perspectives on secondary forest management in tropical humid lowland America. Ambio 31:243–250
Kang BT, Wilson GF (1987) The development of alley cropping as a promising agroforestry technology. In: Steppler HA, Nair PKR (eds) Agroforestry: a decade of development. International Council for Research in Agroforestry, Nairobi, pp 227–244
Kanowski J, Catterall CP, Wardell-Johnson GW, Proctor H, Reis T (2003) Development of forest structure on cleared rainforest land in eastern Australia under different styles of reforestation. For Ecol Manage 183:265–280
Kaothien U, Webster D (2001) Regional development in Thailand: new issues, new responses. In: Edgington DW, Fernandez AL, Hoshino C (eds) New regional development paradigms, vol 2. Greenwood Press, Westport, Connecticut, pp 99–122
Kass DCL, Foletti C, Szott LT, Landaverde R, Nolasco R (1993) Traditional fallow systems of the Americas. Agrofor Syst 23:207–218
Katoppo A (2000) The role of community groups in the environment movement. In: Manning C, VanDiermen P (eds) Indonesia in transition: social aspects of Reformasi and crisis. Zed Books, London, pp 213–219
Keenan RJ, Lamb D, Parrotta J, Kikkawa J (1999) Ecosystem management in tropical timber plantations: satisfying economic, conservation, and social objectives. J Sustain For 9:117–134

Kelty MJ (1992) Comparative productivity of monocultures and mixed-species stands. In: Kelty MJ, Larson BC, Oliver CD (eds) The ecology and silviculture of mixed-species forests. Kluwer, Dordrecht, pp 125–141
Khasa PD, Dancik BP (1997) Managing for biodiversity in tropical forests. J Sustain For 4(1/2):1–31
Kingsbury N, Kellman M (1997) Root mat depths and surface soil chemistry in southeastern Venezuela. J Trop Ecol 13:475–479
Kira T, Shidei T (1967) Primary production and turnover of organic matter in different forest ecosystems of the western Pacific. Jpn J Ecol 17:70–87
Klopfer PF (1959) Environmental determinants of faunal diversity. Am Nat 93:337–342
Kondratyev KYA (1969) Radiation in the atmosphere. International Geophysics Series 12. Academic Press, New York
Korpelainen H, Ådjers G, Kuusipalo J, Nuryanto K, Otsamo A (1995) Profitability of rehabilitation of overlogged dipterocarp forest: a case study from South Kalimantan, Indonesia. For Ecol Manage 79:207–215
Kramer R, van Schaik C, Johnson J (eds) (1997) Last stand protected areas and the defense of tropical biodiversity. Oxford University Press, Oxford
Kusumaatmadja S (2000) Through the crisis and beyond: the evolution of the environment movement. In: Manning C, VanDiermen P (eds) Indonesia in transition: social aspects of Reformasi and crisis. Zed Books, London, pp 203–212
Kuusipalo J, Goran A, Jafarsidik Y, Otsamo A, Tuomela K, Vuokko R (1995) Restoration of natural vegetation in degraded *Imperata cylindrica* grassland: understory development in forest plantations. J Veg Sci 6:205–210
Lal R (1987) Tropical ecology and physical edaphology. Wiley, Chichester
Lamb D (1998) Large-scale ecological restoration of degraded tropical forest lands: the potential role of timber plantations. Restor Ecol 6:271–279
Lamprecht H (1990) Silvicultura en los Trópicos. Deutsche Gesellschaft für Technische Zusammenarbeit (GTZ), Eschborn
Lauer W (1993) Climatology. In: Pancel (ed) Tropical forestry handbook, vol 1. Springer, Berlin Heidelberg New York, pp 96–164
Laurance WF (1998) A crisis in the making: responses of Amazonian forests to land use and climate change. Trends Ecol Evol 13:411–415
Laurance WF (2000) Cut and run: the dramatic rise of transnational logging in the tropics. Trends Ecol Evol 15:433–434
Laurance WF, Fearnside PM (1999) Amazon burning. Trends Ecol Evol 14:457
Laurance WF, Laurance G, Ferreira LV, Rankin de Merona JM, Glascon C, Lovejoy T (1997) Biomass collapse in Amazonian forest fragments. Science 278:117–118
Leopold AC, Andrus R, Finkeldey A, Knowles D (2001) Attempting restoration of wet tropical forests in Costa Rica. For Ecol Manage 142:243–249
Leslie A, Sarre A, Sobral Filho M, bin Buang A (2002) Forest certification and biodiversity. ITTO Newsl. Trop For Update 12:13–15
Levin DA (1976) The chemical defenses of plants to pathogens and herbivores. Annu Rev Ecol Syst 7:121–159
Lewin K (1946) Action research and minority problems. J Soc Issues 2:34–46
Lewis JP (1991) Three levels of floristical variation in the forests of Chaco, Argentina. J Veg Sci 2:125–130
Li W, Han N (2001) Ecotourism management in China's nature reserves. Ambio 30:62–63
Lindenmayer DB, Margules CR, Botkin DB (2000) Indicators of biodiversity for ecologically sustainable forest management. Conserv Biol 14:941–950

Loiselle BA, Dirzo R (2002) Plant–animal interactions and community structure. In: Chazdon RL, Whitmore TC (eds) Foundations of tropical forest biology: classic papers with commentaries. University of Chicago Press, Chicago, pp 269–338

Longman KA, Jenik JJ (1987) Tropical forest and its environment, 2nd edn. Longman, Essex

Lonsdale WM (1988) Predicting the amount of litterfall in forests of the world. Ann Bot 61:319–324

Lopez R (1997) Environmental externalities in traditional agriculture and the impact of trade liberalization: the case of Ghana. J Dev Econ 53:17–39

Louman B, Campos JJ, Schmidt S, Zagt R, Haripersaud P (2002) Los procesos nacionales de certificación forestal y su relación con la investigación forestal. Interacciones entre políticas y manejo forestal, casos de Costa Rica y Guyana. Rev For Centroam 37:41–46

Loumeto JJ, Huttel C (1997) Understorey vegetation in fast-growing eucalypt plantations in the savanna soils in Congo. For Ecol Manage 99:31–36

Loveland TR, Belward AS (1997) The IGBP-DIS global 1 km land cover data set, Discover: first results. Int J Remote Sensing 18:3289–3295

Lovell C, Mandondo A, Moriarty P (2003) The question of scale in integrated natural resource management. In: Campbell BM, Sayer JA (eds) Integrated natural resource management. Linking productivity, the environment and development. CABI, Wallingford, pp 109–136

Lowe PD (1995) The limits to the use of criteria and indicators for sustainable forest management. Comm For Rev 74:343–349

Lugo AE (1988) The future of the forest: ecosystem rehabilitation in the tropics. Environment 30:17–20, 41–45

Lugo AE, Scatena FN (1995) Ecosystem-level properties of the Luquillo Experimental Forest with emphasis on the Tabonuco Forest. In: Lugo AE, Lowe C (eds) Tropical forests: management and ecology. Springer, Berlin Heidelberg New York, pp 59–108

Lugo AE, Figueroa Colon JC, Alayon M (eds) (2002) Big-leaf mahogany: genetics, ecology, and management. Springer, Berlin Heidelberg New York, pp 103–114

MacArthur RH (1969) Patterns of communities in the tropics. Biol J Linnean Soc 1:19–30

MacArthur RH, Wilson EO (1967) The theory of island biogeography. Princeton University Press, Princeton

Mahar D, Schneider R (1994) Incentives for tropical deforestation: some examples from Latin America. In: Brown K, Pearce DW (eds) The causes of tropical deforestation. University Press of British Columbia, Vancouver, pp 159–171

Maiocco DC (1998) Distribución de *Zamia skinneri*, un producto no maderable de los bosques de Centroamérica. Tesis MSc, Centro Agronómico Tropical de Investigación y Enseñanza (CATIE), Turrialba, Costa Rica, 85 pp

Maldonado E, Montagnini F (2004) Carrying capacity of La Tigra National Park, Honduras: can the park be self-sustainable? J Sustain For 19(4)

Mann C (1997) Reseeding the green revolution. Science 277:1038–1043

March JA (2001) Brazilian agrarian reform: potential, problems, and the quest for sustainability. PhD Diss, University of Georgia, Athens, Georgia

Marmillod D, Villalobos R, Robles G (1998) Hacia el manejo sostenible de especies vegetales del bosque con productos no maderables: las experiencias de CATIE en esta década. In: Trabajo presentado al Primer Congreso Latinoamericano IUFRO, Valdivia, Chile

Marquis RJ (1987) Variación en la herbivoría foliar y su importancia selectiva en *Piper aurieianun* (Piperaceae). Rev Biol Trop 35(Suppl 1):133–149
Marquis RJ, Dirzo R (2002) Coevolution. In: Chazdon RL, Whitmore TC (eds) Foundations of tropical forest biology. University of Chicago Press, Chicago, pp 339–347
Matthews JD (1989) Silvicultural systems. Clarendon Press, Oxford
May RM (1972) Will a large complex system be stable? Nature 283:413–414
McCann KS (2000) The diversity-stability debate. Nature 405:228–233
McIntosh RJ (1993) The pulse model: genesis and accommodation of specialization in the Middle Niger. J Afr Hist 34:181–220
McNeely JA, Scherr SJ (2003) Ecoagriculture: strategies to feed the world and save biodiversity. Island Press, Washington, DC, 321 pp
McRae M (1997) Is "good wood" bad for forests? Science 275:1868–1869
Medina E (1995) Physiological ecology of trees and application to forest management. In: Lugo AE, Lowe C (eds) Tropical forests: management and ecology. Ecological Studies 112. Springer, Berlin Heidelberg New York, pp 289–307
Meher-Homji VM (1992) Probable impact of deforestation on hydrological processes. In: Myers N (ed) Tropical forests and climate. Kluwer, Dordrecht, pp 163–174
Mesén F, Cornelius J, Montagnini F (2000) Estrategia de domesticación de *Vochysia guatemalensis* por el CATIE. In: Actas del I Congreso de Investigación. Los retos y propuestas de la investigación en el III milenio, Consejo Nacional de Rectores (CONARE), San José, Costa Rica, 14–16 March, 59 pp
Mesquita RCG (1995) The effect of different proportions of canopy opening on the carbon cycle of a central Amazonian secondary forest. PhD Diss, University of Georgia, Athens, Georgia
Michelsen A, Lisanework N, Friis I, Holst N (1996) Comparison of understory vegetation and soil fertility in plantations and adjacent natural forests in the Ethiopian highlands. J Appl Ecol 33:627–642
Michon G, Mary F, Bompard J (1989) Multistoreyed agroforestry garden system in West Sumatra, Indonesia. In: Nair PKR (ed) Agroforestry systems in the tropics. Kluwer, Dordrecht, pp 243–268
Molino J, Sabatier M (2001) Tree diversity in tropic rain forests: a validation of the intermediate disturbance hypothesis. Science 294:1702–1704
Monk CD (1966) An ecological significance of evergreenness. Ecology 47:504–505
Montagnini F (1992) Sistemas Agroforestales. Principios y Aplicaciones en los Trópicos, 2nd edn. OTS/CATIE, San José, Costa Rica, 622 pp
Montagnini F (2000) Accumulation in aboveground biomass and soil storage of mineral nutrients in pure and mixed plantations in a humid tropical lowland. For Ecol Manage 134:257–270
Montagnini F (2001) Strategies for the recovery of degraded ecosystems: experiences from Latin America. Interciencia 26:498–503
Montagnini F (2002) Tropical plantations with native trees: their function in ecosystem restoration. In: Reddy MV (ed) Management of tropical plantation-forests and their soil litter system. Litter, biota and soil-nutrient dynamics. Science Publishers, Enfield, New Hampshire, pp 73–94
Montagnini F, Jordan CF (2002) Ciclaje de nutrients en bosques lluviosos neotropicales. In: Guariguata MR, Kattan G (eds) Ecología y Conservación de Bosques Lluviosos Neotropicales. Ediciones LUR, Cartago, Costa Rica, pp 167–191
Montagnini F, Mendelsohn R (1996) Managing forest fallows: improving the economics of swidden agriculture. Ambio 26:118–123

Montagnini F, Muñiz-Miret N (1999) Vegetation and soils of tidal floodplains of the Amazon estuary: a comparison of varzea and terra firme forests in Pará, Brazil. J Trop For Sci 11:420–437

Montagnini F, Nair PK (2004) Carbon sequestration: an under-exploited environmental benefit of agroforestry systems. Agrofor Syst 61:281–295

Montagnini F, Porras C (1998) Evaluating the role of plantations as carbon sinks: an example of an integrative approach from the humid tropics. Environ Manage 22:459–470

Montagnini F, Sancho F (1990) Impacts of native trees on tropical soils: a study in the Atlantic lowlands of Costa Rica. Ambio 19:386–390

Montagnini F, Sancho F (1994) Aboveground biomass and nutrients in young plantations of four indigenous tree species: implications for site nutrient conservation. J Sustain For 1:115–139

Montagnini F, Sancho F, Ramstad K (1993) Litter fall, litter decomposition and the use of mulch of four indigenous tree species in the Atlantic lowlands of Costa Rica. Agrofor Syst 23:39–61

Montagnini F, Fanzeres A, da Vinha SG (1994) Studies on restoration ecology in the Atlantic Forest region of Bahia, Brazil. Interciencia 19:323–330

Montagnini F, González E, Rheingans R, Porras C (1995a) Mixed and pure forest plantations in the humid neotropics: a comparison of early growth, pest damage and establishment costs. Commonwealth For Rev 74:306–314

Montagnini F, Fanzeres A, da Vinha SG (1995b) The potential of twenty indigenous tree species for reforestation and soil restoration in the Atlantic Forest region of Bahia. J Appl Ecol 32:841–856

Montagnini F, Eibl B, Grance L, Maiocco D, Nozzi D (1997) Enrichment planting in overexploited subtropical forests of the Paranaense region of Misiones, Argentina. For Ecol Manage 99:237–246

Montagnini F, Eibl B, Woodward C, Szczipanski L, Ríos R (1998) Tree regeneration and species diversity following conventional and uniform spacing methods of selective cutting in a subtropical humid forest reserve. Biotropica 30:349–361

Montagnini F, Jordan CF, Matta Machado R (2000) Nutrient cycling and nutrient use efficiency in agroforestry systems. In: Ashton MS, Montagnini F (eds) The silvicultural basis for agroforestry systems. CRC Press, Boca Raton, pp 131–160

Montagnini F, Finegan B, Delgado D, Eibl B, Szczipanski L, Zamora N (2001) Can timber production be compatible with conservation of forest biodiversity? Two case studies of plant biodiversity in managed neotropical forests. J Sustainable For 12(1/2):37–60

Montagnini F, Campos JJ, Cornelius J, Finegan B, Guariguata M, Marmillod D, Mesén F, Ugalde L (2002) Environmentally-friendly forestry systems in Central America. Bois For Trop 272(2):33–44

Montagnini F, Ugalde L, Navarro C (2003) Growth characteristics of some native tree species used in silvopastoral systems in the humid lowlands of Costa Rica. Agrofor Syst 59:163–170

Montero MM, Kanninen M (2002) Biomasa y carbono en plantaciones de *Terminalia amazonia* (Gmel.). Excell en la zona Sur de Costa Rica. Rev For Centroam 39/40:50–55

Mooney HA, Field C, Vázquez Yanez C (1984) Photosynthetic characteristics of wet tropical forest plants. In: Medina E, Mooney HA, Vázquez-Yanes C (eds) Physiological ecology of plants of the wet tropics. W Junk, The Hague, pp 113–128

Moran EF (1981) Developing the Amazon. Indiana University Press, Bloomington

Moran EF (1996) Deforestation in the Brasilian Amazon. In: Sponsel NE, Headland TE, Bailey RD (eds) Tropical deforestation: the human dimension. Columbia University Press, New York, pp 149–164

Muñiz-Miret N, Vamos R, Hiraoka M, Montagnini F, Mendelsohn R (1996) The economic value of managing açai (*Euterpe oleracea* Mart.) in floodplain lands of the Amazon estuary, Pará, Brazil. For Ecol Manage 87:163–173

Muntingh H (1997) Indonesia: mega-rice project and "free logging" in Central Kalimantan. http://forests.org/archive/indomalay/megarice.htm

Murphy PG, Lugo AE, Murphy AJ, Nepstad DC (1995) The dry forests of Puerto Rico's south coast. In: Lugo AE, Lowe C (eds) Tropical forests: management and ecology. Ecological Studies 112. Springer, Berlin Heidelberg New York, pp 178–209

Myers N (1984) The primary source: tropical forests and our future. Norton, New York

Myers N (1988) Tropical deforestation and climatic change. Environ Conserv 15:293–298

Myers N (1992) Tropical forests and climate. Reprinted from Climatic Change 19 (1991). Kluwer, Dordrecht, 265 pp

Myers N (1994) Population and biodiversity. In: Graham Smith E (ed) Population: the complex reality. The Royal Society, London, pp 117–136

Myers N (1996) The world's forests: problems and potential. Environ Conserv 23:156–168

Myers N (1997) Our forestry prospect: the past recycled or a surprise-rich future? The Environmentalist 17:233–247

Myers N (2003) Biodiversity hotspots revisited. BioScience 53:916–917

Myers N, Mittermeier RA, Mittermeier CG, da Fonseca GAB, Kent J (2000) Biodiversity hotspots for conservation priorities. Nature 403:853–858

Nadkarni N (1981) Canopy roots: convergent evolution in rainforest nutrient cycles. Science 214:1023–1024

Naeem S, Thompson ILJ, Lawler SP, Lawton JH, Woodfin RM (1994) Declining biodiversity can alter performance of ecosystems. Nature 368:734–736

Nair PKR (1989) Agroforestry defined. In: Nair PKR (ed) Agroforestry systems in the tropics. Kluwer, Dordrecht, pp 13–18

Nair PKR (1990) The prospects and promise of agroforestry in the tropics: a review of technical and socioeconomic information with special emphasis to Africa. Report to the World Bank, Washington, DC, 121 pp

Nasi R, Wunder S, Campos AJJ (2002) Forest ecosystem services: can they pay our way out of deforestation? Discussion paper prepared for the Global Environmental Facility for the Forestry Roundtable, in conjunction with the UN Forum on Forests II, Costa Rica, 11 March

Nations JD (1988) The Lacandon Maya. In: Denslow JS, Padoch C (eds) People of the tropical rain forest. University of California Press, Berkeley, pp 86–88

Navarro C, Wilson J, Gillies A, Hernandez M (2002) A new Mesoamerican collection of big-leaf mahogany. In: Lugo A, Figueroa JC, Mildred Alayón C (eds) Big leaf mahogany: genetics, ecology and management. Ecological Studies 159. Springer, Berlin Heidelberg New York, pp 103–117

Navarro C, Montagnini F, Hernández G (2004) Genetic variability of *Cedrela odorata* Linnaeus: results of early performance of provenances and progenies from Mesoamerica grown in association with coffee. For Ecol Manage 192(2/3):217–227

Nemani RR, Keeling CC, Hashimoto H, Jolly WM, Piper SC, Tucker CJ, Myneni RB, Running SW (2003) Climate-driven increases in global net primary production from 1982 to 1999. Science 300:1560–1562

Nepstad D, Uhl C, Adilson Serrão E (1990) Surmounting barriers to forest regeneration in abandoned, highly degraded pastures: a case study from Paragominas, Pará, Brazil. In: Anderson AB (ed) Alternatives to deforestation: steps toward sustainable use of the Amazon rain forest. Columbia University Press, New York, pp 215–229

Nepstad D, Uhl C, Serrao EAS (1991) Recuperation of a degraded Amazonian landscape: forest recovery and agricultural restoration. Ambio 20:248–255

Nepstad D, DeCarvalho CR, Davidson EA, Jipp PH, Lefebre PA, Negreiros GH, da Silva ED, Stone TA, Trumbore SE, Vieira S (1995) The deep-soil link between water and carbon cycles of Amazonian forests and pastures. Nature 372:666–667

Nepstad D, McGrath D, Alencar A, Barros AC, Carvalho G, Santilli M, Vera Diaz M del C (2002) Frontier governance in Amazonia. Science 295:629–630

Newton AC, Baker P, Ramnarine S, Mesén JF, Leakey RRB (1993) The mahogany shoot borer: prospects for control. For Ecol Manage 57:301–328

Noble IR, Dirzo R (1997) Forests as human-dominated ecosystems. Science 277:522–525

Nobre CA, Sellers PJ, Shukla J (1991) Amazonian deforestation and regional climate change. J Climate 4:957–988

Nortcliff S, Thornes JB (1978) Water and cation movement in a tropical rainforest environment. Acta Amazonica 8:245–258

Norton-Griffiths M (1998) The economics of wildlife conservation policy in Kenya. In: Milner-Gulland EJ, Mace R (eds) Conservation of biological resources. Blackwell, Oxford

Nye PH, Greenland DJ (1960) The soil under shifting cultivation. Tech Comm 51. Commonwealth Bureau of Soils, Commonwealth Agricultural Bureaux, Farnham Royal, Buckinghamshire

Nykvist N (1997) Total distribution of plant nutrients in a tropical rainforest ecosystem, Sabah, Malaysia. Ambio 26:152–157

O'Dowd DJ, Green PT, Lake PS (2003) Invasional "meltdown" on an oceanic island. Ecol Lett 6:812–817

Odum EP (1969) The strategy of ecosystem development. Science 164:262–270

Odum HT (1970) The El Verde study area and the rain forest systems of Puerto Rico. In: Odum HT, Pigeon RF (eds) A tropical rain forest: a study of irradiation and ecology at El Verde, Puerto Rico. Division of Technical Information, US Atomic Energy Commission, Washington, DC, pp B3–B32

Oldeman RAA, van Dijk J (1991) Diagnosis of the temperament of tropical rain forest trees. In: Gómez-Pompa A, Whitmore TC, Hadley M (eds) Rain forest regeneration and management. Man and the Biosphere Series vol 6. UNESCO, Paris, and The Parthenon Publishing Group, Carnforth, UK, pp 21–65

Olson JS (1963) Energy storage and the balance of producers and decomposers in ecological systems. Ecology 44:322–332

Orians G, Apple JL, Billings R, Fournier L, Gilbert L, McNab B, Sarukhan J, Smith N, Stiles G (1974) Tropical population ecology. In: Farnworth EG, Golley FB (eds) Fragile ecosystems: evaluation of research and applications in the Neotropics. Springer, Berlin Heidelberg New York, pp 5–65

Otsamo A, Hadi TS, Kurniati L, Vuokko R (1999) Early performance of 12 *Acacia crassicarpa* provenances on an *Imperata cylindrica* dominated grassland in South Kalimantan, Indonesia. J Trop For Sci 11:36–46

Pagiola S (2002) Paying for water services in Central America: learning from Costa Rica. In: Pagiola S, Bishop J, Landell-Mills N (eds) Selling forest environmental services. Market-based mechanisms for conservation and development. Earthscan, London, pp 37–61

Park TK (1992) Early trends toward class stratification: chaos, common property, and flood recession agriculture. Am Anthropol 94:90–117

Parrotta JA (1992) The role of plantation forests in rehabilitating degraded tropical ecosystems. Agric Ecosyst Environ 41:115–133

Parrotta JA (1999) Productivity, nutrient cycling, and succession in single- and mixed-species plantations of *Casuarina equisetifolia*, *Eucalyptus robusta*, and *Leucaena leucocephala* in Puerto Rico. For Ecol Manage 124:45–77

Parrotta JA, Knowles OH, Wunderlee Jr JM (1997) Development of floristic diversity in 10-year-old restoration forests on a bauxite mined site in Amazonia. For Ecol Manage 99:21–42

Pearce D, Brown K (1994) Saving the world's tropical forests. In: Brown K, Pearce DW (eds) The causes of tropical deforestation. University of British Columbia Press, Vancouver, pp 2–26

Pearce D, Putz FE, Vanclay J (2003) Sustainable forestry in the tropics: panacea or folly? For Ecol Manage 172:229–247

Pereira Jr R, Zweede J, Asner GP, Keller M (2002) Forest canopy damage and recovery in reduced-impact and conventional selective logging in eastern Pará, Brazil. For Ecol Manage 168:77–89

Peres CA, Baider C, Zuidema PA, Wadt LHO, Kainer KA, Gomes-Silva DAP, Salomao RP, Simoes LL, Franciosi ERN, Valverde EFC, Gribel R, Shepard Jr GH, Kanashiro M, Coventry P, Yu DW, Watkinson AW, Freckleton RP (2003) Demographic threats to the sustainability of Brazil nut exploitation. Science 302:2112–2114

Pérez CLD, Kanninen M (2002) Wood specific gravity and aboveground biomass of *Bombacopsis quinata* plantations in Costa Rica. For Ecol Manage 165:1–9

Peters CM, Gentry AH, Mendelsohn RO (1989) Valuation of an Amazonian rainforest. Nature 339:655–656

Phothitai M (1992) Taungya in Thailand: perspective of the Forest Industry Organization. In: Jordan CF, Gajaseni J, Watanabe H (eds) Taungya: forest plantations with agriculture in Southeast Asia. CAB International, Wallingford, pp 87–94

Pianka ER (1966) Latitudinal gradients in species diversity. A review of concepts. Am Nat 100:33–46

Pimentel D, Stachow U, Takacs DA, Brubaker HW, Dumans AR, Meaney JJ, O'Neil JAS, Onsi DE, Corzilius DB (1992) Conserving biological diversity in agricultural/forestry systems. BioScience 42:354–362

Piotto D, Montagnini F, Ugalde L, Kanninen M (2003a) Performance of forest plantations in small and medium sized farms in the Atlantic lowlands of Costa Rica. For Ecol Manage 175:195–204

Piotto D, Montagnini F, Ugalde L, Kanninen M (2003b) Growth and effects of thinning of mixed and pure plantations with native trees in humid tropical Costa Rica. For Ecol Manage 177:427–439

Piotto D, Montagnini F, Kanninen M, Ugalde L, Viquez E (2004a) Forest plantations in Costa Rica and Nicaragua: performance of species and preferences of farmers. J Sustain For 18(4):59–77

Piotto D, Víquez E, Montagnini F, Kanninen M (2004b) Pure and mixed forest plantations with native species of the dry tropics of Costa Rica: a comparison of growth and productivity. For Ecol Manage 190:359–372

Posey DA (1982) The keepers of the forest. Garden 6:18–24

Potter C, Davidson E, Nepstad D, de Carvalho C (2001) Ecosystem modeling and dynamic effects of deforestation on trace gas fluxes in Amazon tropical forests. For Ecol Manage 152:97–117

Potter RL, Jordan CF, Guedes RM, Batmanian GJ, Han XG (1991) Assessment of a phosphorus fractionation method for soils: problems for further investigation. In: Crossley DA, Coleman DC, Hendrix PF, Cheng W, Wright DH, Beare MH, Edwards CA (eds) Modern techniques in soil ecology. Elsevier, Amsterdam, pp 443–463

Powers JS, Haggar JP, Fisher RF (1997) The effect of overstory composition on understory woody regeneration and species richness in 7-yr-old plantations in Costa Rica. For Ecol Manage 99:43–54

Prance GT, Beentje H, Dransfield J, Johns R (2000) The tropical flora remains undercollected. Ann Mo Bot Gard 87:67–71

Preeyagrysorn O (1992) Taungya in Thailand: perspective of the Royal Forestry Department. In: Jordan CF, Gajaseni J, Watanabe H (eds) Taungya: forest plantations with agriculture in Southeast Asia. CAB International, Wallingford, pp 95–100

PRORENA (2003) The native species reforestation project (PRORENA) strategic plan 2003–2008. Document 20. Center for Tropical Forest Science (CTFS), Smithsonian Tropical Research Institute (STRI), and Tropical Resources Institute at the Yale School of Forestry and Environmental Studies, New Haven, Connecticut

Ramos JM, del Amo S (1992) Enrichment planting in a tropical secondary forest in Veracruz, Mexico. For Ecol Manage 54:289–304

Raven PH (1988) Our diminishing tropical forests. In: Wilson EO, Peters FM (eds) Biodiversity. National Academy Press, Washington, DC, pp 119–122

Reich PB, Uhl C, Walters MB, Prugh L, Ellsworth D (2004) Leaf demography and phenology in Amazonian rain forest: a census of 40,000 leaves of 23 tree species. Ecol Monogr 74:3–35

Reid JW, Rice RE (1997) Assessing natural forest management as a tool for tropical forest conservation. Ambio 26:382–386

Repetto R (1988) Overview. In: Repetto R, Gillis M (eds) Public policies and the misuse of forest resources. Cambridge University Press, Cambridge, pp 1–41

Repetto R (1992) Accounting for environmental assets. Sci Am June:94–100

Rhoades RE, Booth R (1982) Farmer-back-to-farmer: a model for generating acceptable technology. Agric Admin 11:127–137

Richards PW (1952) The tropical rain forest. Cambridge University Press, Cambridge

Richards PW (1996) The tropical rainforest. An ecological study, 2nd edn. Cambridge University Press, London

Ricker M, Siebe C, Sánchez BS, Shimada K, Larson BC, Martínez-Ramos M, Montagnini F (2000) Optimizing seedling management: *Pouteria sapota*, *Diospyros digyna*, and *Cedrela odorata* in a Mexican rainforest. For Ecol Manage 139:63–77

Robles G, Ocampo R, Marmillod D (1997) Incorporación de una especie no maderable en un sistema silvicultural diversificado: el caso de *Zamia skinneri*. In: Centro Agronómico Tropical de Investigación y Enseñanza (CATIE), Actas de la III Semana Científica, Turrialba, Costa Rica, 3–5 Feb, pp 133–138

Rosenberg T (2002) Globalization. New York Times Magazine, 18 Aug, Sect 6, pp 28–75

Rosenzweig ML (1995) Species diversity in space and time. Cambridge University Press, Cambridge

Roosevelt AC, Lima da Costa M, Lopes Machado C, Michab M, Mercier N, Valladas H, Feathers J, Barnett W, Imazio da Silveira M, Henderson A, Sliva J, Chernoff B, Reese DS, Holman JA, Toth N, Schick K (1996) Paleoindian cave dwellers in the Amazon: the peopling of the Americas. Science 272:373–384

Rosset P, Mittal A (2001) The paradox of plenty. Wall Street Journal, 7 Jan, p A27

Salati E, Nobre CA (1992) Possible climatic impacts of tropical deforestation. In: Myers N (ed) Tropical forests and climate. Kluwer, Dordrecht, pp 177–196

Salati E, Vose PB (1984) Amazon basin: a system in equilibrium. Science 225:129–138

Sanchez PA (1976) Properties and management of soils in the tropics. Wiley, New York

Santana R, Montagnini F, Louman B, Villalobos R, Gómez M (2002) Productos de bosques secundarios del Sur de Nicaragua con potencial para la elaboración de artesanías de Masaya. Rev For Centroam 38:85–90

Sawyer D (1990) The future of deforestation in Amazonia: a socioeconomical and political analysis. In: Anderson AB (ed) Alternatives to deforestation. Columbia University Press, New York, pp 265–274

Saxena KG, Rao KS, Sen KK, Maikhuri RK, Semwal RL (2003) Integrated natural resource management: approaches and lessons from the Himalaya. In: Campbell BM, Sayer JA (eds) Integrated natural resource management. Linking productivity, the environment and development. CABI, Wallingford, pp 211–225

Sayer JA, Campbell BM (2003) Research to integrate productivity enhancement, environmental protection, and human development. In: Campbell BM, Sayer JA (eds) Integrated natural resource management. Linking productivity, the environment and development. CABI, Wallingford, pp 1–14

Schmidt RC (1991) Tropical rain forest management: a status report. In: Gomez-Pompa A, Whitmore TC, Hadley M (eds) Rain forest regeneration and management. Man and the Biosphere Series. UNESCO, Paris, pp 181–207

Schneider RR (1995) Government and the economy of the Amazon frontier. World Bank Environment Pap 11. World Bank, Washington, DC

Schroeder P (1994) Carbon storage benefits of agroforestry systems. Agrofor Syst 27:89–97

Schulze PC, Leighton M, Peart DR (1994) Enrichment planting in selectively logged rain forest: a combined ecological and economic analysis. Ecol Appl 4:581–592

Scott GAJ (1978) Grassland development in the Gran Pajonal of eastern Peru. PhD Diss, Department of Geography, University of Hawaii, Manoa, Honolulu

Sedjo RA (1999) The potential of high-yield plantation forestry for meeting timber needs. New For 17:339–359

Shepherd D, Montagnini F (2001) Carbon sequestration potential in mixed and pure tree plantations in the humid tropics. J Trop For Sci 13(3):450–459

Simpson GG (1964) Species density of North American recent mammals. Syst Zool 13:57–73

Sips P (1993) Management of tropical secondary rain forests in Latin America: today's challenge, tomorrow's accomplished fact!? National Reference Center for Nature, Forests and Landscape, Ministry of Agriculture, Nature Management and Fisheries, Wageningen, The Netherlands, 72 pp

Sist P, Nolan T, Bertault JG, Dykstra D (1998) Harvesting intensity versus sustainability in Indonesia. For Ecol Manage 108:251–260

Sist P, Sheil D, Kartawinata K, Priyadi H (2003) Reduced-impact logging in Indonesian Borneo: some results confirming the need for new silvicultural prescriptions. For Ecol Manage 179:415–427

Smith DM (1986) The practice of silviculture, 8th edn. Wiley, New York, 527 pp

Smouts MC (2003) Tropical forests, international jungle: the underside of global ecopolitics. Palgrave Macmillan, New York

Soemarwoto O (1987) Homegardens: a traditional agroforestry system with a promising future. In: Steppler HA, Nair PKR (eds) Agroforestry: a decade of development. International Council for Research in Agroforestry, Nairobi, pp 157–171

Specht RL (1988) Origin and evolution of terrestrial plant communities in the wet-dry tropics of Australia. In: Kitching RL (ed) The ecology of Australia's wet tropics. Proc Ecol Soc Aust 15:19–30

Stadtmüller T (1987) Cloud forests in the humid tropics. United Nations University, Tokyo

Stanley W, Montagnini F (1999) Biomass and nutrient accumulation in pure and mixed plantations of indigenous tree species grown on poor soils in the humid tropics of Costa Rica. For Ecol Manage 113:91–103

Stark NM, Jordan CF (1978) Nutrient retention by the root mat of an Amazonian rainforest. Ecology 59:434–437

Stevens CJ, Dise NB, Mountford JO, Gowing DJ (2004) Impact of nitrogen deposition on the species richness of grasslands. Science 303:1876–1879

Stockdale EA, Lampkin NH, Hovi M, Keatinge R, Lennartsson EKM, Macdonald DW, Padel S, Tattersall FH, Wolfe MS, Watson CA (2001) Agronomic and environmental implications of organic farming systems. Adv Agron 70:261–327

Stokstad E (2003) "Pristine" forest teemed with people. Science 301:1645–1646

Stone RD, D'Andrea C (2001) Tropical forests and the human spirit. University of California Press, Berkeley

Stouffer PC, Borges SH (2001) Conservation recommendations for understory birds in Amazonian forest fragments and second-growth areas. In: Bierregaard RO, Gascon C, Lovejoy TE, Mesquita RCG (eds) Lessons from Amazonia: the ecology and conservation of a fragmented forest. Yale University Press, New Haven, pp 248–261

Stradling DJ (1978) Food and feeding habits of ants. In: Brian MV (ed) Production ecology of ants and termites. Cambridge University Press, Cambridge, pp 81–106

Subler S, Uhl C (1990) Japanese agroforestry in Amazonia: a case study in Tomé Açu, Brazil. In: Anderson AB (ed) Alternatives to deforestation: steps toward sustainable use of the Amazon Rain Forest. Columbia University Press, New York, pp 152–166

Swift MJ, Heal OW, Anderson JM (1979) Decomposition in terrestrial ecosystems. University of California Press, Berkeley

Takeda S (1992) Origins of Taungya. In: Jordan CF, Gajaseni J, Watanabe H (eds) Taungya: forest plantations with agriculture in Southeast Asia. CAB International, Wallingford, pp 9–17

Tanner EVJ, Vitousek PM, Cuevas E (1998) Experimental investigation of nutrient limitation of forest growth on wet tropical mountains. Ecology 79:10–22

Terborgh J (1973) On the notion of favorableness in plant ecology. Am Nat 107:481–501

Terborgh J (1999) Requiem for nature. Island Press, Washington, DC

Thadani R (2001) International non-timber forest product issues. J Sustainable For 13:5–23

Thoms CA, Betters DR (1998) The potential for ecosystem management in Mexico's forest ejidos. For Ecol Manage 103:149–157

Tilman D (1987) Secondary succession and the pattern of plant dominance along experimental nitrogen gradients. Ecol Monogr 57:189–214

Tilman D (2000) Causes, consequences and ethics of biodiversity. Nature 405:208–211

Trenbath BR (1986) Resource use by intercrops. In: Francis CA (ed) Multiple cropping systems. Macmillan, New York, pp 57–81

Tricart J (1972) The landforms of the humid tropics: forests and savannas. Longman, London

Trist ABAE, Higgin G, Murray H, Pollock A (1963) Organizational choice. Tavistock, London

Turner IM (2001) The ecology of trees in the tropical rain forests. Cambridge Tropical Biology Series. Cambridge University Press, Cambridge

Uhl C, Clark H, Clark K, Maquirino P (1982) Successional patterns associated with slash-and-burn agriculture in the upper Rio Negro of the Amazon Basin. Biotropica 14:249–254

Uhl C, Nepstad D, Buschbacher R, Clark K, Kauffman K, Subler S (1990) Studies of ecosystem response to natural and anthropogenic disturbances provide guidelines for designing sustainable land-use systems in Amazonia. In: Anderson AB (ed) Alternatives to deforestation: steps toward sustainable use of the Amazon rain forest. Columbia University Press, New York, pp 24–42

Uhl C, Verissimo A, Mattos MM, Brandino Z, Vieira ICG (1991) Social, economic, and ecological consequences of selective logging in an Amazonian frontier: the case of Tialandia. For Ecol Manage 46:243–273

UNESCO (1978) Tropical forest ecosystems. A state-of-knowledge report prepared by UNESCO/UNEP. FAO, Paris

Upton C, Bass S (1996) The certification handbook. International Institute for the Environment and Development, London

Urquhart GR, Skole DL, Chomentowski WH, Barber CP (1998) Tropical deforestation. NASA Facts. FS-1998-11-120-GSFC. National Aeronautics and Space Administration, Goddard Space Flight Center, Greenbelt, Maryland

USDA (1990) Draft amended land and resource management plan, Caribbean National Forest/Luquillo Experimental Forest. USDA Southern Forest Experiment Station, Rio Piedras, Puerto Rico

USGS (2004) Global land cover characteristics data. http://edcdaac.usgs.gov/glcc.globdoc2_0. html

Vanclay JK (1992) Species richness and productive forest management. In: Miller FR, Adam KL (eds) Wise management of tropical forests. Oxford Forestry Institute, Oxford, pp 1–9

Van der Hammen T (1974) The Pleistocene changes of vegetation and climate in tropical South America. J Biogeogr 1:3–26

Vandermeer JH (1990) Intercropping. In: Carroll CR, Vandermeer JH, Rosset P (eds) Agroecology. McGraw-Hill, New York, pp 481–516

Van der Pijl L (1972) Principles of dispersal in higher plants. Springer, Berlin Heidelberg New York

Van Wambeke A (1992) Soils of the tropics: properties and appraisal. McGraw-Hill, New York

Varmola MI, Carle JB (2002) The importance of hardwood plantations in the tropics and subtropics. Int For Rev 4:110–121

Vázquez-Yanes C, Orozco-Segovia A (1990) Seed dormancy in the tropical rain forest. In: Bawa KS, Hadley H (eds) Reproductive ecology of tropical forest plants. Man and the Biosphere Series. UNESCO, Paris, pp 247–259

Verissimo A, Barreto P, Mattos M, Tarifa R, Uhl C (1992) Logging impacts and prospects for sustainable forest management in an old Amazonian frontier: the case of Paragominas. For Ecol Manage 55:169–199

Vincent LW (2002) RIL and sustainable tropical forest management. Int Soc Trop For News 23:9

Vitousek P (1981) Clear-cutting and the nitrogen cycle. In: Clark FE, Rosswall T (eds) Terrestrial nitrogen cycles. Ecol Bull (Stockh) 33:631–642
Vitousek P (1982) Nutrient cycling and nutrient use efficiency. Am Nat 119:553–572
Vitousek P (1984) Litterfall, nutrient cycling, and nutrient limitation in tropical forests. Ecology 65:285–298
Vitousek PM, Reiners WA (1975) Ecosystem succession and nutrient retention: a hypothesis. BioScience 25:376–381
Wade R (2001) Global inequality. The Economist, 28 April, pp 72–74
Wadsworth FH (1997) Forest production in tropical America. USDA Agricultural Handbook 710. USDA, Washington, DC
Wadsworth F (1999) Editorial: low impact logging: what is it? Int Soc Trop For News 20(2):5
Waggener T (2001) Role of forest plantations as substitutes for natural forests in wood supply – lessons learned from the Asia-Pacific region. Forest plantations thematic paper series. FAO, Rome
Wallace AR (1878) Tropical nature and other essays. Macmillan, London, pp 65–68
Walter H (1971) Ecology of tropical and subtropical vegetation. Oliver and Boyd, Edinburgh
Wang D, Bormann FH, Lugo AE, Bowden RE (1991) Comparison of nutrient-use efficiency and biomass production in five tropical tree taxa. For Ecol Manage 46:1–21
Weaver PL (1987) Enrichment planting in tropical America. In: Figueroa Colón JC, Wadsworth FH, Branham S (eds) Management of the forests of tropical America: prospects and technologies. International Institute of Tropical Forestry, USDA Forest Service, and University of Puerto Rico, Río Piedras, pp 258–278
Weaver PL (1993) Secondary forest management. In: Parrotta JA, Kanashiro M (eds) Management and rehabilitation of degraded lands and secondary forests in Amazonia. International Institute of Tropical Forestry, USDA Forest Service, and UNESCO Man and the Biosphere Program, Río Piedras, Puerto Rico, pp 117–128
Weaver PL (1995) The Colorado and dwarf forests of Puerto Rico's Luquillo mountains. In: Lugo AE, Lowe C (eds) Tropical forests: management and ecology. Ecological Studies 112. Springer, Berlin Heidelberg New York, pp 109–141
Webb WL, Lauenroth WK, Szarek SR, Kinerson RS (1983) Primary production and abiotic controls in forests, grasslands, and desert ecosystems in the United States. Ecology 64:134–151
Went FW, Stark N (1968) Mycorrhiza. BioScience 18:1035–1039
Westoby JC (1962) Forest industries in the attack on underdevelopment in the state of food and agriculture. FAO, Rome
Whitmore JL (1999) The social and environmental importance of forest plantations with emphasis on Latin America. J Trop For Sci 11:255–269
Whitmore TC (1984) Tropical rain forests of the Far East, 2nd edn. Clarendon Press, Oxford
Whitmore TC (1990) An introduction to tropical rain forests. Clarendon Press, Oxford
Whitmore TC (1991) Tropical rain forest dynamics and its implications for management. In: Gomez-Pompa A, Whitmore TC, Hadley M (eds) Rain forest regeneration and management. UNESCO, Paris, and The Parthenon Publishing Group, Carnforth, UK, pp 67–89
Whitmore TC (1997) Tropical forest disturbance, disappearance, and species loss. In: Laurance WF, Bierregaard RO (eds) Tropical forest remnants: ecology, manage-

ment, and conservation of fragmented communities. University of Chicago Press, Chicago, pp 3–12
Whitmore TC (1998) An introduction to tropical rain forests, 2nd edn. Oxford University Press, Oxford, 282 pp
Whittaker RH (1975) Communities and ecosystems. Macmillan, New York
Whittaker RH, Likens CE (1975) The Biosphere and Man. Ecological Studies 14. Springer, Berlin Heidelberg New York, pp 305–328
Whittaker RH, Marks PL (1975) Methods of assessing terrestrial productivity. In: Lieth H, Whittaker RH (eds) Primary productivity of the biosphere. Ecological Studies 14. Springer, Berlin Heidelberg New York, pp 55–118
Wilkie DS (1988) Hunters and farmers of the African forest. In: Denslow JS, Padoch C (eds) People of the tropical rain forest. University of California Press, Berkeley, pp 111–126
Wilkie DS (1996) Logging in the Congo: implications for indigenous foragers and farmers. In: Sponsel LE, Headland TN, Bailey RC (eds) Tropical deforestation: the human dimension, pp 230–247. Columbia University Press, New York
Wilkie DS, Sidle JG, Boundzanga GC (1992) Mechanized logging, market hunting, and a bank loan in the Congo. Conserv Biol 6:570–580
Willis KJ, Whittaker RJ (2002) Species diversity – scale matters. Science 295:1245–1248
Wilson EO (1992) The diversity of life. Harvard University Press, Cambridge, 424 pp
Wood TG (1978) Food and feeding habits of termites. In: Brian MV (ed) Production ecology of ants and termites. Cambridge University Press, Cambridge, pp 55–80
World Bank (1988) Indonesia: the transmigration program in perspective. World Bank, Washington, DC
World Resources Institute (2000) World Resources 2000–2001. People and ecosystems. The fraying web of life. World Resources Institute, Washington, DC, 389 pp
World Resources Institute (2004) Land area classification by ecosystem type. www.earthtrends.wri.org
Wormald TJ (1992) Mixed and pure forest plantations in the tropics and subtropics. FAO Forestry Pap 103. FAO, Rome, 152 pp
Wyatt-Smith J, Panton WP (1963) Manual of Malayan silviculture for inland forest. Malayan Forest Record 23. Forest Research Institute, Kuala Lumpur, 350 pp
Yap SK, Chan HT (1990) Phenological behaviour of some *Shorea* species in peninsular Malaysia. In: Bawa KS, Hadley M (eds) Reproductive ecology of tropical forest plants. Man and the Biosphere Series. UNESCO, Paris, pp 21–35
Young A (1997) Agroforestry for soil management, 2nd edn. CAB International, Wallingford
Zech W (1993) Geology and soils. In: Pancel L (ed) Tropical forestry handbook. Springer, Berlin Heidelberg New York, pp 1–93
Zeng N (2003) Drought in the Sahel. Science 302:999–1000
Zinke PJ, Stangenberger AG, Post WM, Enamuel WR, Olson JS (1984) Worldwide organic soil carbon and nitrogen data. Publ 2212, Oak Ridge National Laboratory, Environmental Sciences Division, US Department of Energy, Washington, DC

Subject Index

M

P

Printing: Krips bv, Meppel
Binding: Litges & Dopf, Heppenheim